Data Deduplication for High Performance Storage System

Dan Feng

Data Deduplication for High Performance Storage System

 Springer

Dan Feng
School of Computer Science and Technology
Huazhong University of Science and
Technology
Wuhan, China

ISBN 978-981-19-0114-0   ISBN 978-981-19-0112-6   (eBook)
https://doi.org/10.1007/978-981-19-0112-6

This Springer imprint is published by the registered company Springer Nature Singapore Pte Ltd.
The registered company address is: 152 Beach Road, #21-01/04 Gateway East, Singapore 189721, Singapore

Preface

According to deduplication studies conducted by Microsoft and EMC, about 50% and 85% of the data in their production primary and secondary storage systems, respectively, are redundant and could be removed by the deduplication technology. Data deduplication, an efficient approach to data reduction, has gained increasing attention and popularity due to the explosive growth of digital data. Deduplication has been commercially available since 2004 in a dedicated backup appliance, and multiple vendors have created competing products. Since that time, deduplication has become a standard feature on numerous storage systems. Nowadays, deduplication provides many benefits not only for storage systems but also for other application scenarios, such as reducing duplicate I/Os for primary storage, avoiding transmitting duplicate data for network environments, and extending lifetime of emerging non-volatile storage devices.

In this Synthesis Lecture, we provide a detailed overview of data deduplication technologies focusing on storage systems. We begin with a discussion of requirements for deduplication, and then present overview of deduplication systems. It then describes the problems facing each phase of data deduplication workflow and the state-of-the-art technologies to solve the problems.

The organization of this Synthesis Lecture is as follows:

- Chapter 1 describes the background and development of backup systems. Moreover, the redundancy distribution and deduplication-based backup systems are also discussed.
- Chapter 2 discusses the requirements for data deduplication technologies and presents overview of deduplication systems.
- Chapter 3 describes state-of-the-art chunking algorithms, analysis, and our proposed chunking algorithms.
- Chapter 4 describes existing indexing schemes and our proposed indexing scheme.
- Chapter 5 describes state-of-the-art schemes to reduce fragmentation and our proposed rewriting algorithm.

- Chapter 6 introduces the potential attacks and requirements of security in data deduplication systems, state-of-the-art schemes and our proposed scheme to ensure data confidentiality, simplify key management and reduce computation over-heads.
- Chapter 7 presents two fast post-deduplication delta compression schemes.
- Chapter 8 presents an open-source platform for deduplication and the selection of the deduplication technologies. There have been many publications about data deduplication, while no solution can do best in all metrics. The goal of the platform is to find some reasonable solutions that have sustained backup performance and a suitable trade-off between deduplication ratio, memory footprint, and restore performance.

This Synthesis Lecture is the outcome of a series of academic papers presented by the authors at the 2019 and 2015 USENIX Conference on File and Storage Technologies (FAST); the 2011, 2014, and 2016 USENIX Annual Technical Conference (ATC); the 2015 IEEE Conference on Mass Storage Systems and Technologies (MSST); the 2015 IEEE International Conference on Computer Communications (INFOCOM); the 2014 IEEE Data Compression Conference (DCC); the 2014 ACM International Symposium on Computer Performance, Modeling, Measurements and Evaluation (Performance); the 2011 IEEE International Parallel and Distributed Processing Symposium (IPDPS); the 2016 Proceedings of the IEEE; the 2016 IEEE Transactions on Parallel and Distributed Systems; the 2015, 2016, 2017, and 2020 IEEE Transactions on Computers. Many thanks for a lot of pioneering research done by the doctoral students: Wen Xia, Min Fu, Yukun Zhou, Yucheng Zhang, and You Chen.

The Synthesis Lecture is intended for a broad audience: graduate students, academic and industry researchers, and practitioners. We believe that this lecture will provide a complete view for the reader to learn about the data deduplication technologies and systems.

Wuhan, China   Dan Feng

Contents

Abbreviations

AE	Asymmetric Extremum
CABdedup	Causality-based Deduplication
CaFTL	Content-aware Flash Translation Layer
CAP	Capping Algorithm
CBR	Context-Based Rewriting algorithm
CDC	Content-Defined Chunking
CFL	Chunk Fragmentation Level
DDFS	Data Domain File System
DER	Deduplication Elimination Ratio
DS	Deduplication Server
ESG	Enterprise Storage Group
FSC	Fixed-Size Chunking
GCC	GNU Compiler Collection
GDB	GNU symbolic debugger
HAR	History-Aware Rewriting algorithm
HPDS	High-Performance Deduplication System
IPC	Instructions Per Cycle
LAN	Local Area Network
NC	Normalized Chunking
OPT	Optimal Cache
RAID	Redundant Arrays of Independent Disks
RAM	Random Access Memory
SSD	Solid State Disk
TRE	Traffic Redundancy Elimination
WAN	Wide Area Network

Chapter 1
Deduplication: Beginning from Data Backup System

Abstract Data backup is an important application to prevent data loss caused by a disaster. But it will generate a lot of redundant data to increase the cost. In this chapter, we will describe the definitions, schemes, and key technologies of data backup, and introduce the data deduplication research originally from the data backup system. Section 1.1 describes the background of backup. Section 1.2 describes the generation of redundant data and the advantages of deduplication. Section 1.3 describes a basic architecture of a deduplication-based backup system.

1.1 Background

1.1.1 Development of Backup System

With the rapid development of information society, the operations of people's everyday life and business are surrounded by increasingly digital information and information systems. Jim Gray (a Turing Award winner) has come up with a rule of thumb. The amount of data produced every 18 months in a network environment equals the sum of the amount of data ever created. While information systems provide fast service decisions and easy management, people also have to face the danger of data loss [1]. Therefore, protecting the integrity and security of data is of particular importance to enterprises. The loss of enterprise business data will threaten its normal business operations and threaten the competitive edge and market reputation of enterprises. According to IDC, 55% of companies in the United States had suffered a disaster from 2000 to 2010. About 29% of the fortunate companies fail in the next two years due to data loss. Finally, only 16% of the companies survived. There are many types of disasters, including hardware failures, human error, software defects, computer virus attacks, and natural disasters. It can also result in billions of dollars in financial losses. Data backup [2] is the last line of defense for data protection in order to quickly recover data when the data crashes and improve data availability.

Data backup refers to the original copy of the data to generate a copy of the original data. When the original data is damaged or lost, the use of the backup data has been restored to recover the original data. Data backup has experienced a four-stage development :stand-alone backup, Local Area Network (LAN) centralized backup, remote backup, and cloud backup. First, stand-alone backup refers to the data on the machine backup to the same machine on a different partition or a different folder under the same partition. This backup method is very simple, which does not need to consume additional hardware devices and network resources. However, the disaster tolerance is poor. When the machine is destroyed, the data will not be restored. Second, LAN backup refers to the data backup to connect to machines in the same LAN. This backup is generally operated by a professional backup manager. They utilize a centralized and unified backup and management way to backup data from a number of machines to other backup a number of machines in an idle time. When system data crashes, the data is restored by a professional backup manager. However, this LAN backup can only tolerate the data within the same LAN. If there is a disaster locally, this backup cannot do anything. Third, remote backup refers to establish a full-time backup center in the different places.

The data are transferred to the remote backup center via a high-speed dedicated network. The cost of this backup solution is high. Generally, to reduce the cost of dedicated network, thus the system chooses the same city or region as a full-time backup center. For many large enterprises or government departments that require high security and reliability, they usually choose a full-time and dedicated backup center. Compared with LAN backup, remote backup with a dedicated backup center can tolerate disasters at the site, for example, fire accident. However, this method cannot tolerate disasters throughout the urban areas, such as earthquakes and tornadoes. Fourth, cloud backup is a data backup service that transfers data to backup data center via Wide Area Network (WAN). Users can freely choose to buy the required backup service. Cloud backup is a service-oriented backup service, and the main purpose is to reduce the overall cost of data backup. LAN centralized backup has a poor disaster tolerance. The cost is too high for remote backup with a dedicated backup center. And hence most small and medium enterprises tend to use cloud backup services to meet their backup needs. Finally, hot standby is a high-availability backup system based on multiple servers, for example, two servers. According to the switching modes, hot standby is classified by active-standby and active-active. The active-standby mode means that a server is in an active state of service, and the other server is in the standby state of the service. The active-active mode means two different services are mutually active and standby (Active-Standby and Standby-Active) on two servers respectively. Once there is a server downtime, another server can provide services immediately. Therefore, we have to back up the status and data between the two servers.

1.1.2 Features of Backup System

The backup data needs to be stored, and probably should be organized to a degree. The organization could include a computerized index, catalog, or relational database.

Different approaches have different advantages. The key part of the backup model is the backup rotation scheme.

There are three common backup strategies in backup systems, namely full backup, incremental backup, and differential backup.

First, full backup refers to perform a complete backup for all the current data. A repository of this type contains complete system images taken at one or more specific points in time. The backup data has nothing to do with the history of backup versions. Users can recover data by finding the corresponding full backup sets, and the recovery speed is fast.

Second, incremental backup is only required to back up the data that was modified since the last backup version. An incremental style repository aims to make it more feasible to store backups from more points in time by organizing the data into increments of change between points in time. This eliminates the need to store duplicate copies of unchanged data; with full backups a lot of the data will be unchanged from what has been backed up previously. Compared with the full backup, incremental backup requires to upload a little amount of data. Incremental backups need to focus on the history of backup versions. Additionally, some backup systems can reorganize the repository to synthesize full backups from a series of incremental backups. At the time of data recovery, users need to find out the most recent full backup and all the incremental backup data sets between this full backup and incremental backup. The recovery speed is very slow.

Third, differential backup refers to backup data that has been modified since the last full backup. Each differential backup saves the data that has changed since the last full backup. It has the advantage that only a maximum of two data sets are needed to restore the data. One disadvantage, compared to the incremental backup method, is that as time from the last full backup and thus the accumulated changes in data increases, so does the time to perform the differential backup. Therefore, differential backup needs more backup data than incremental backup, but less than the full backup. As the same with incremental backup, differential backup also needs to focus on the history of backup version. However, differential backup is concerned about the last full backup. Incremental backup focuses on the last backup, which can be full backup, incremental backup, and differential backup. In data recovery, differential backup only needs to find out the datasets of the most recent full backup. The recovery speed of differential backup is faster than incremental backup, but slower than full backup.

1.2   Deduplication in Backup Systems

1.2.1   Large-Scale Redundant Data

The amount of digital data in the world is growing explosively, as evidenced in part by the significant increase in the estimated amount of data generated in 2010 and 2011 from 1.2 zettabytes to 1.8 zettabytes, respectively [4, 5], and the predicted

Table 1.1 Summary of recently published analyses on redundant data reduction ratios of large-scale real-world datasets by industry

Institutes	Workloads	Reduction ratio
Microsoft	857 desktop computers	About 42%
	15 MS file servers	15–90%
EMC	Over 10,000 production storage system	69–93%
	6 large storage datasets	85–97%
Symantec	40,000 backup systems	More than 88%

amount of data to be produced in 2020 is 44 zettabytes [6]. As a result, how to efficiently store and transfer such large volumes of digital data is a challenging problem. However, the workload studies conducted by Microsoft, EMC, and Symantec suggest the existence of a large amount of redundant data in their production primary and secondary storage systems [7–11], as shown in Table 1.1 [62]. There are very few changes between versions of backup data. Thus there will be lots of duplicate and similar data.

Therefore, data deduplication, a space and bandwidth efficient technology that prevents redundant data from being stored in storage devices and transmitted over the networks, is one of the most important methods to tackle this challenge. Deduplication has been commercially available since 2004 in a dedicated backup appliance [12], and multiple vendors have created competing products [13–16]. Since that time, deduplication has become a standard feature on numerous storage systems. It has been widely used in memory system to load more data, solid-state storage system to reduce the number of writes and improve SSD life, etc.

1.2.2   Why Deduplication?

With the rapid growth of storage capacity in recent years, large number of service-driven data center have already appeared. Data backup and recovery face the challenges of time and space. In the space dimension, the backup system needs to store more and more backup data. How to use the limited storage resources to efficiently store petabytes or even exabytes of data becomes a critical problem. In the time dimension, remote backup and cloud backup have been rapidly promoted and popularized. Data backup needs to be done quickly under limited network bandwidth, which reduces the impact to other upper-level applications. When disaster strikes, data recovery needs to be done as quickly as possible and reduce economic losses. In the era of information, we face serious challenges both in time and space. The traditional backup technologies (full backup, incremental backup and differential backup) cannot meet the requirements. In particular, they cannot meet the requirements of remote backup and cloud backup with limited space and time constraints. Therefore, data deduplication is an efficient compression technology to solve these problems.

Data deduplication can eliminate large-scale redundant data and reduce storage cost. For example, if we copy a 10 MB file to 100 users, we need to use a 1 GB storage system. If all users back up their data each week, we need an additional 1 GB storage space per week. After one year, a total of 52 GB of storage space is required. However, we only need a 10 MB storage space to backup these data if we utilize data deduplication. Thus, data deduplication becomes a very challenging and attractive topic in a large-scale backup and archiving system. According to the research of Enterprise Storage Group (ESG), the newly added data that reference the previously stored data occupy more than half of the total data. In addition, the amount of data increases by 68% per year. The referenced data contain a large number of duplicate and stable data content. The data are stored for a long time. According to the study, there are large amount of duplicate data in all aspects of information processing and data storage. Duplicate data are mainly from some important applications, for example, file systems, file synchronization, email attachments, html documents, web objects, and operating systems. These duplicate data may be uploaded to backup systems to prevent disasters.

Traditional backup methods cannot identify the redundant data in the backup workloads. It will waste lots of network bandwidth and storage space. And it also reduces the utilization of storage space for backup and archive systems. The study shows that the compression rate can reach more than 20 in the data backup systems. Remote backup and cloud backup systems both face challenges in space and time, but data deduplication brings a new revolution. Data deduplication not only reduces storage space, but also reduces resource overheads and maintenance costs. In addition, data deduplication can reduce the amount of data that needs to be transferred. By this way, data deduplication can save network bandwidth and accelerate the process of backup and recovery.

1.3   Deduplication-Based Backup System

A typical backup system consists of three entities, which provides data backup and restore services. In Fig. 1.1, it consists of backup server, storage server, and clients. Backup server and storage server are service providers and can be located in the same LAN environment and Wide Area Networks (WAN). Clients connect to the server over the WAN and enjoy the backup service. The command flow is separated from the data flow in this system.

- Backup Server: is a control center of the system management. Backup server is responsible for the following tasks: handling service request, creating a task, scheduling, and resource management. Backup server needs to manage lots of resources, such as user's resources, storage resources, backup strategies, metadata, and system log information. All of the information is stored in the database.
- Storage Server: is a data processing center of the system. Storage server receives control command from the backup server. When the storage server receives data

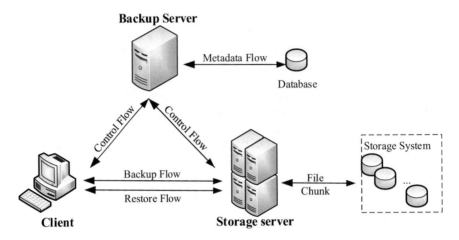

Fig. 1.1 Basic architecture of backup system

from clients, it will organize data and build index effectively. The information of data and index is stored on the storage devices. When receiving the request of recovery, the storage server will read data from the storage devices and send them to the clients.

- Client: Users leverages the client to perform the operations of backup and restore. The client is the starter of the service and controlled by the backup server. The client reads data from the disk locally and sends them to the storage server for backup service. The client receives data from the storage server when users perform data restore requests. Users can customize the backup strategy through the client.

In backup systems, there are many backed-up versions of the same dataset stored at the backup destination storage system due to multiple full and incremental backups. Except for the initial full backups, the dataset backed up by a client each time is evolved from its previous backed-up versions with data modifications. As a result, it is the modified data, rather than the unmodified data that is already stored at the storage server by previous backed-up versions, that is required to be transmitted each time. The same is true during the data restores. Each restore operation takes place after data corruptions and needs to restore the corrupted dataset to a previous backed-up version stored at the backup destination. Just as the dataset backed up each time, the corrupted dataset requiring restore is evolved from its previous backup versions with data modifications done in users' local computers. Thus the data that has not been modified after backups has no need to be transmitted from the backup destination. Therefore, during either data backup or restore operations, it is possible to reduce the amount of data transmitted over network by removing from transmission the unmodified data shared between the current dataset being processed (i.e., dataset to be backed up or restored each time) and its previous backed-up versions.

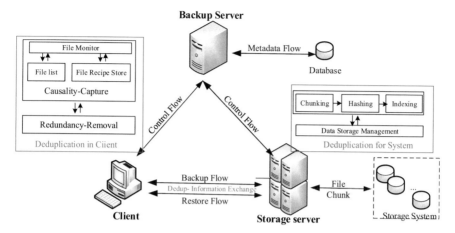

Fig. 1.2 Deduplication-based backup system

Figure 1.2 shows a deduplication-based backup system. There are two schemes to implement data deduplication, which are global deduplication on the storage server side and optional local deduplication on the client side. The global deduplication on the storage server can be used independently or in combination with the local deduplication on the client in a backup system.

As shown in Fig. 1.2, global data deduplication is implemented on the storage server. When the data stream reaches the storage server, the first stage of data deduplication is data chunking, that is, the data stream is partitioned according to a certain block size. Then hashing the data block to generate a short fingerprint that uniquely identifies the block. The matching hash fingerprints mean that the content they represent is likely to be the same as each other, thereby identifying duplicate data blocks while avoiding byte-by-byte comparisons. All fingerprints are managed together. When there is a new data block, it is determined whether there is duplicate data by indexing the fingerprints. A very important part of the deduplication system is data storage management, including fingerprints management and data chunking management. To facilitate high-speed fingerprint indexing, fingerprints are usually organized as key-value stores (such as hash tables), so only one lookup is needed to identify duplicate data blocks. Each index entry is a key-value pair, where the key is a fingerprint and the value is the location of the corresponding data block. The container storage for data blocks is a storage subsystem with a log structure. While eliminating duplicate blocks, gather unique blocks into a fixed-size (e.g., 4 MB) container storage.

As shown in Fig. 1.2, the optional local deduplication is implemented on the client side. It is composed of two functional modules: the Causality-Capture module and Redundancy-Removal module. The former, consisting of File Monitor, File List, and File Recipe Store, is responsible for capturing the causal relationships among different files, while the latter is responsible for removing the unmodified data for each backup/restore operation with the help of the captured causality information by the former. Of the components of the Causality-Capture module, File Monitor is a daemon process that works at the file system level to keep track of some specific file operations, including file rename, file create, file delete, and file-content modification, File List is responsible for logging these file operations and File Recipe Store is responsible for saving the file recipes (i.e., the fingerprints of data chunks) of the backed-up files. These three components collectively capture the causal relationships among the different files in multiple backups and restores. The identification and removal of the unmodified data become the key to improving both the backup and restore performances for a client.

1.4   Concluding Remarks

Data backup is a basic service and has been widely used in lots of applications. In this chapter, we present the background of data backup. After that, we show the classification and methods of backup. Due to data backup services, there are large amount of redundant data to be stored in the storage systems. And hence, data deduplication is an efficient and system-level compression approach to eliminate duplicate data. Finally, we describe the system architecture of deduplication-based backup systems.

Chapter 2
Overview of Data Deduplication

Abstract Data deduplication is a lossless data compression technology. Deduplication eliminates redundant data by keeping only one physical copy that could be referenced by other duplicate data (copies), which reduces storage cost and network bandwidth. Section 2.1 presents the principle and methods of data deduplication on different types. Section 2.2 describes the workflow and procedure of data deduplication. Section 2.3 describes some application scenarios of data deduplication. Section 2.4 presents the challenges of data deduplication.

2.1 The Principle and Methods of Data Deduplication

Deduplication can be implemented at different granularities: a file, a chunk, a byte, and a bit. The finer the data granularity, the more redundant data is deleted. Meantime, the finer the data granularity, the more computational resource are consumed. There are lots of deduplication methods in backup systems according to the data granularity, range, time, and place.

2.1.1 File-Level and Chunk-Level Deduplication

The granularity can be divided into file-level, chunk-level, byte-level, and bit-level. Recently, file-level and chunk-level deduplication are widely used.

File-level deduplication refers to the reduction of data storage by deleting files with the same content that saves storage space. It uses a hash function to generate the hash value of each file from its data content. If two data have the same hash value, they are considered as identical. The duplicate data only keeps one physical copy, which reduces storage cost. However, file-level deduplication cannot identify and eliminate similar files, thus the compression ratio is limited. Many existing systems use file-level deduplication, for example, SIS[16], FarSite [17], and EMC Center [18].

© Springer Nature Singapore Pte Ltd. 2022
D. Feng, *Data Deduplication for High Performance Storage System*,
https://doi.org/10.1007/978-981-19-0112-6_2

Chunk-level deduplication eliminates duplicate data on chunks, which is more flexible and reduces more storage overheads. It divides a file stream into fixed-size or variable-size chunks. The average chunk size is 4KB-8KB. It utilizes a hash function (e.g., SHA-1/SHA-256) to calculate the hash value of each chunk. The hash value is referred to as "fingerprint." Fingerprint is used to uniquely identify each chunk. The two chunks that have the same fingerprint are regarded as the duplicate chunks. Compared with file-level deduplication, chunk-level deduplication not only deletes duplicate files but also eliminates the duplicate chunks in the similar files. Generally, chunk-level deduplication is efficient and can get a higher deduplication ratio, which now is the most widely used deduplication method. Chunk-level deduplication is used on Data Domain File System (DDFS). However, chunk-level deduplication is more fine-grained, thus also consumes more computational resource.

There are also some byte-level and bit-level deduplication methods, except for chunk-level and file-level deduplication technologies. For example, ExGrid combines file-level and chunk-level deduplication, and leverages the redundancy and similarity to find and eliminate duplicate data.

2.1.2   Local and Global Deduplication

According to the search range of duplicate data, data deduplication can be classified into local and global deduplication. Local deduplication is a method that only deleting redundant data on the same client, on the same machine, or on the same storage node. Local deduplication cannot eliminate redundant data across multiple node. However, the search range of local deduplication is limited. From the perspective of compression ratio, global deduplication can achieve good compression efficiency. Global deduplication needs to search duplicate data on multiple nodes. From the perspective computational cost, the process of global deduplication introduces large overheads. Many companies make a tradeoff between local and global deduplication, for example, IBM and EMC.

2.1.3   Online and Offline Deduplication

According to the time of deduplication, data deduplication consists of online and offline deduplication (post-processing deduplication). Online deduplication refers to deleting duplicate data before the data reaches the storage device, and the storage device only stores only unique nonduplicate data blocks. The lookup and deletion operation of online deduplication occur on the critical path, which will affect the performance of storage systems. Improper implementation ways can seriously affect the data write performance of storage systems. To solve this problem, offline deduplication utilizes a disk buffer. All pending data are firstly stored on the disk buffer. After all data have been written to the disk, some data in the disk buffer will be reread in some idle time of the system for deduplication. Offline deduplication is a post-processing scheme. Offline deduplication does not affect the storage performance of the upper application at run time, but it needs extra storage space as disk buffer. Therefore, for the real-time critical applications with high storage performance, offline deduplication is more suitable than online deduplication. However, data backup is not a real-time application in the current backup systems. Data backup is a non-real-time application, therefore it often uses online deduplication to eliminate redundant data in backup systems. The file systems require high storage performance, thus offline deduplication is more appropriate.

2.1.4 Source-Based and Target-Based Deduplication

According to the location of deduplication, deduplication contains the sender and the destination. Data deduplication is classified to source-based and target-based deduplication. The source is the data sender, for example, the client. The targets are the data receiver and the storage backend, for example, the server. Source-based deduplication is a deduplication scheme that deletes redundant data before sending data. And hence, duplicate data do not need to be transferred and stored. It is suitable for low bandwidth network system to transfer data via WAN, for example, cloud storage and cloud backup. This way can greatly reduce the amount of data and accelerate the process of transmission. However, the process of searching and deleting on duplicate data are performed at client, which will consume computational resources at source. Thus it will affect the performance of applications on the source machine. Target-based deduplication is a method that performs deduplication to search and delete redundant data at server. Therefore, all the overheads of deduplication will be brought to the target side. If the transmission bandwidth is high, target-based deduplication is more appropriate than the source-based deduplication.

2.2 Basic Workflow of Data Deduplication

Data deduplication technologies can be divided into two categories: file level and chunk level. File-level deduplication detects and removes duplicate files, while chunk-level deduplication divides file into chunks and eliminates identical chunks. The chunk-level deduplication can detect more redundant data because nonduplicate files may have identical chunk which can be eliminated by chunk-level deduplication but cannot be removed by file-level deduplication. It worth noting that file-level deduplication was proposed earlier, but that technology was subsequently overshadowed by chunk-level deduplication due to the latter's better compression performance. In this chapter, we focus on chunk-level deduplication.

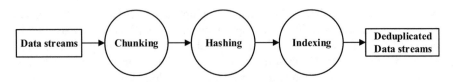

Fig. 2.1 Basic workflow of deduplication technology

2.2.1   Workflow of chunk-level Deduplication

Chunk-level deduplication (hereafter data deduplication) follows the workflow of chunking, fingerprint, and indexing, as shown in Fig. 2.1.

Chunking. The first stage of the basic workflow of date deduplication is chunking. During chunking stage, data streams are divided into multiple chunks for deduplication. The size of the chunks has a significant impact on the deduplication efficiency, which reflects the amount of removed redundant data. Smaller chunks contribute to more detected redundant data, thus higher deduplication efficiency. However, to simplify the process of duplication identification, the chunks will be hashed to generate short fingerprints which uniquely identify the chunks. Dividing the data stream into smaller chunks means generating more fingerprints, which complicates the fingerprint management. EMC indicates that 8KB strikes a good tradeoff between deduplication efficiency and fingerprint cost.

Bell Labs propose to divide the data stream into fixed-size chunks by applying fixed-size chunk algorithm. Fixed-size chunk algorithm suffers from low deduplication efficiency due to boundaries-shifting problem. Given two files A and B, and the file B is generated by inserting one byte at the beginning of the file A. Dividing two files into chunks by applying fixed-size chunking algorithm, we will find that compared with the chunk boundaries in file A, all chunk boundaries in file B are shifted by one byte and no duplicate chunks can be detected. To solve the problem, MIT Laboratory proposed the content-defined chunking algorithm, which declares the chunk boundaries depending on local content. If the local content is not changed, the chunk boundaries will not be shifted.

Hashing. A simple method to check whether two chunks are identical is comparing the contents of the chunks byte-by-byte. This method is not fit for large-scale storage system, because determining whether a chunk is identical to one of a large amount of processed chunks using this method needs to maintain all processed chunks in RAM which will quickly overflow the RAM. The hashing method simplifies the process of duplicate identification. During this stage, a cryptographic hash function (e.g., SHA-1, SHA-256) will be calculated for each chunk to generate a short fingerprint. The matched fingerprints mean that their represented contents are, with high probability, identical to each other, thus identifying duplicate chunks while avoiding a byte-by-byte comparison. Here the fingerprints refer to a family of cryptographic hash functions that has the key property of being practically infeasible to (1) find two different messages with the same hash and (2) generate a message from a given hash.

Table 2.1 Hash collision probability analysis of SHA-1, SHA-256, and SHA-512 with difference sizes of unique data and with an average chunk size of 8KB

Size of unique data	SHA-1 160 bits	SHA-256 256 bits	SHA-512 512 bits
1GB (2^{30} B)	10^{-38}	10^{-67}	10^{-144}
1TB (2^{40} B)	10^{-32}	10^{-61}	10^{-138}
1PB (2^{50} B)	10^{-26}	10^{-55}	10^{-132}
1EB (2^{60} B)	10^{-20}	10^{-49}	10^{-126}
1ZB (2^{70} B)	10^{-14}	10^{-43}	10^{-120}
1YB (2^{80} B)	10^{-8}	10^{-37}	10^{-114}

According to the "birthday paradox," the collision probability of a given SHA-1 pair can be calculated as follows:

$$\text{Hash calculation : fingerprint} = \text{Hc(content) (fingerprint length of } m \text{ bits)}$$

$$\text{Hash collision : } p \leq \frac{n(n-1)}{2} \times \frac{1}{2^m} \ (n \text{ is the number of the chunks})$$

Table 2.1 shows the hash collision probability according to the above-mentioned equation with varying amounts of unique data[62]. The probability of a hash collision when data deduplication is carried out in an EB-scale storage system, based on the average chunk size of 8KB and fingerprints of SHA-1, is smaller than 10^{-20} (about 2^{-67}). In contrast, in computer systems, the probability of a hard disk drive error is about 10^{-12}–10^{-15}, which is much higher than the aforementioned probability of SHA-1-fingerprint collisions in data deduplication. Consequently, SHA-1 has become the most wildly used fingerprinting algorithm for data deduplication because most existing approaches, such as LBFS, Venti, and DDFS, consider the hash collision probability to be sufficiently small to be ignored when applying deduplication in a PB-scale storage system. More recently, stronger hash algorithms, such as SHA-256, have been considered for fingerprinting in some data deduplication systems, such as ZFS and Dropbox, to further reduce the risk of hash collision.

Indexing. After chunking and hashing stages, chunk fingerprints are indexed to help determine the duplicate and nonduplicate chunks, which is a critical stage for the deduplication process. During indexing stage, each chunk will be checked to see if it is duplicate by comparing its fingerprint with the fingerprints of the previous processed chunks. A fingerprint index that includes the fingerprints of all processed nonduplicate chunks is required in this stage. To facilitate fingerprint matching, the fingerprint index is often organized as a key-value store, such as hash table, thus only requiring one lookup to identify a duplicate chunk. Each index entry is a key-value pair with the key being a fingerprint and the value being the location of the corresponding chunk. Figure 2.2 shows an example of fingerprint index.

Specifically, to check if a new chunk is duplicate or not, a lookup request to the finger-print index is required. If there exists an identical fingerprint in the fingerprint index, the new chunk is duplicate. Otherwise, the chunk is nonduplicate and its fingerprint will be inserted into the fingerprint index for matching.

As the size of the data continues to grow from TB to PB and even EB scale, the total size of the fingerprints representing the user data grows quickly. It is impractical

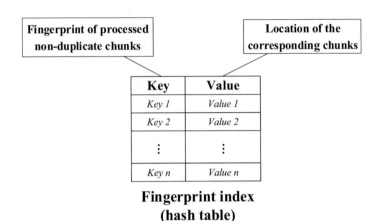

Fig. 2.2 An example of fingerprint index. Assume that the index is organized as a hash table

to keep such a large fingerprint index in RAM and a disk-based index with one seek per incoming chunk is far too slow. Data Domain solves the problem by leveraging the locality among multiple data streams. Symantec exploits the inherent locality of the data stream with a progressive sampled indexing approach to further reduce memory overhead for fingerprint indexing.

2.2.2 Procedure of Data Deduplication

While data deduplication has become widely implemented, we focus our discussion on deduplication storage system because it has the most complicated workflow and its workflow includes most of processes of the workflow of the deduplication technologies applied to the other scenarios. Figure 2.3 depicts an example of deduplication storage system. It has three components, namely fingerprint index, recipe store, and write-evict unit store.

Fingerprint Index. Fingerprint index is used to identify duplicate chunks. It maps fingerprints of stored chunks to their physical locations, as described in Figure 2.2. However, with the explosive growth of data volume, the total number of fingerprints and thus the size of their index increase exponentially, quickly overflowing the RAM capacity of the system. For example, to index a unique dataset of 1PB and assuming an average chunk size of 8KB, about 2.5TB worth of SHA-1 fingerprints (160 bits each chunk) will be generated. To solve the problem, a fingerprint cache is maintained in RAM that holds hot fingerprints to boost duplication identification.

Recipe Store. Recipe store manages recipes that describe the logical fingerprint sequences of processed files or data streams. A recipe is used to reconstruct a file or data stream during restore. It is a list that includes the metadata of a file or data stream and a sequence of fingerprints of the chunks consisting of the file or data stream and the physical location of the chunks, as depicted in Figure 2.4. During

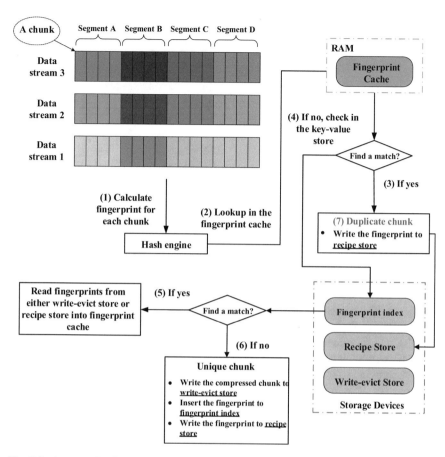

Fig. 2.3 An example of deduplication storage system

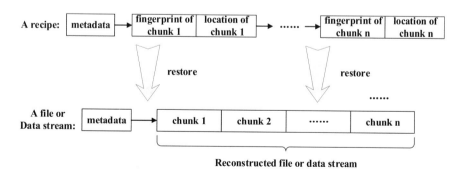

Fig. 2.4 An example of restoring a file or data stream from a recipe

Fig. 2.5 Description of
write-evict unit

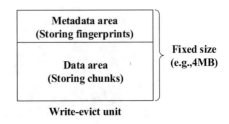

Write-evict unit

restore, the chunks consisting of the file or data stream can be found according to the fingerprints and the location information in the recipe, and the sequences of finger-prints will be replaced by the contents of the corresponding chunks, thus reconstructing the file or data stream.

Write-evict Unit Store. Write-evict unit store is a log-structure storage structure. While duplicate chunks are eliminated, unique chunks are aggregated into fixed-sized (e.g., 4 MB) write-evict unit. Figure 2.5 shows an example of a write-evict unit. It consists of two areas, namely, metadata area and data area. The data area stores chunks and the metadata area stores the fingerprints of the chunks in data area.

Figure 2.3 also shows an example of deduplication procedure in storage system. At the top left, we have three sample data streams. Each data stream is divided into chunks, and four consecutive chunks constitute a segment in our case (assuming a simplest segmenting approach). Each chunk is processed in the following steps: (1) Hash engine calculates the SHA-1 digest for the chunk as its unique identifica-tion, namely fingerprint. (2) Look up the fingerprint in the in-RAM fingerprint cache. (3) If we find a match, jump to step (7). (4) Otherwise, look up the fingerprint in the fingerprint index. (5) If we find a match, a fingerprint read procedure is invoked Jump to step (7). (6) Otherwise, the chunk is unique. We write the chunk to write-evict store, insert the fingerprint to fingerprint index, and write the fingerprint to recipe store. Jump to step (1) to process next chunk. (7) The chunk is duplicate. We eliminate the chunk and write its fingerprint to recipe store. Jump to step (1) to process next chunk.

2.3   Application Scenarios of Data Deduplication

Data deduplication schemes have been widely applied in many scenarios of com-puter systems. The typical applications will be introduced, and how they incorporate and benefit from deduplication. We introduce the different application scenarios, goals, and representative works in Table 2.2. By means of eliminating duplicate data, data deduplication provides many benefits, for example, saving storage space for secondary storage, reducing duplicate I/Os for primary storage, avoiding transmit-ting duplicate data for network environments, and extending lifetime of emerging

Table 2.2 Summary of the application scenarios, goals and benefits, and representative works of data deduplication

Scenarios	Goals and benefits	Representative works
Secondary storage	Saving storage space, and time	Venti , DDFS
Primary storage	Reducing storage cost and disk I/O	Microsoft, iDedup
Cloud Storage	Reducing storage cost and time	Cumulus, Dropbox
Solid State Storage	Extending lifetime	CAFTL, R-Dedup
Network	Reducing time for WAN	Spring et al, EndRE
Virtual machines	Saving memory and migration time	VMware, Vmflock

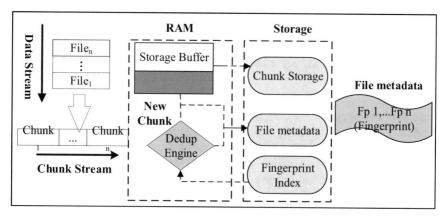

Fig. 2.6 An example of the deduplication architecture for secondary (backup/archive) storage systems

nonvolatile storage devices. Then we describe the application of data deduplication in different scenarios as follows.

2.3.1 Deduplication for Secondary Storage

There is an abundance of duplicates in secondary storage systems, such as backup and archive storage. Based on this observation, Data Domain, a storage company and now a part of EMC, argues that "disk-based deduplication storage has emerged as the new-generation storage system for enterprise data protection to replace tape libraries." In fact, the use of data deduplication in such systems has been shown to achieve a data reduction factor of about 5–40, leading to significant savings in storage space and corresponding hardware costs. More recently, post-deduplication delta compression has been used to compress the nonduplicate but similar data chunks as a complementary approach to data deduplication. Such post-deduplication schemes are shown to achieve an extra data reduction factor of 2–5 on top of data deduplication but adding additional computation and I/O overheads.

Figure 2.6 presents an example of the deduplication architecture for secondary (backup/archive) storage system. It splits a data stream into multiple chunks and each chunk is identified by a hash signature. A dedup engine is used to lookup the fingerprints in fingerprint the index for duplicate checking. The fingerprint index is too large to store in the memory for large-scale storage systems. Thus the dedup engine reads a part of hot fingerprints into RAM to accelerate the lookup operation and insert the new fingerprints into index. New chunks will be stored in the storage buffer. The chunks will be written into chunk storage when the storage buffer is full. A file metadata that consists of the fingerprint sequences of the backup is prepared for future data recovery.

2.3.2   Deduplication for Primary Storage

Primary storage data deduplication not only reduces the storage space requirement but also eliminates duplicate I/Os, which helps improve the disk I/O performance for primary storage. More recently, several open-source file systems, such as ZFS and OpenDedupe have incorporated deduplication for better storage performance. Figure 2.7 gives an example of the deduplication architecture for primary storage system. The deduplication engine consists of a file system filter driver and a set of background jobs. The deduplication filter redirects the file system interactions (read, write, etc.). The background jobs are designed to run in parallel with the primary server I/O workload. The deduplication engine maintains the chunks and their metadata in large chunk store, which enable fast sequential I/O access.

Procedure for reading a file. For a nondeduplicated file, the read operation is executed regularly from the file system. For a fully deduplicated file, a file read request will be partitioned into a sequence of chunk read requests through the file table. The table is composed of a list of chunk fingerprints and their logical addresses. Then the chunk fingerprint is parsed to extract the index of its physical address. Finally, the read request is reassembled from the contents of the chunks. For partially deduplicated files, the system maintains an extra bitmap for each file to keep

Fig. 2.7 An example of the deduplication architecture for primary storage systems

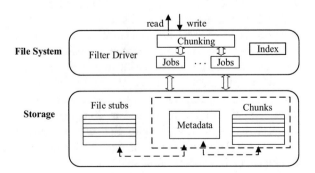

a record of the nondeduplicated regions. The bitmaps are stored in the original file contents that are marked for a "sparse" stream.

Procedure for writing a file. The operation of file writes does not change the content, it simply overwrites some bytes in the associated sparse stream. After content overwriting, some chunks may hold data that is old with respect to this file. The deduplication filter has the metadata that contains a sequence of chunk fingerprints and addresses. Thus, the system can rebuild a file from the allocated sparse streams and ranges backed up by chunks.

2.3.3   Deduplication for Cloud Storage

Cloud storage has emerged as an important storage platform for computer systems in recent years. Since the limited network bandwidth in the underlying wide area network (WAN) imposes the main performance bottleneck of cloud storage, data deduplication can help accelerate the data synchronization between client and cloud by identifying the unmodified data. In the meantime, data deduplication also helps reduce the storage overhead in the cloud side. Currently, DropBox, SkyDrive (now called OneDrive), Google Drive, etc., incorporate data deduplication to provide better cloud storage service.

Figure 2.8 describes an example of the deduplication architecture for cloud storage system. This architecture is also suitable for remote backup or cloud backup systems with C/S model. The architecture consists of client and server. The input data are read to RAM and divided into chunks via chunking algorithms (Fixed-size or Contend-defined Chunking algorithm). For each chunk, the client generates chunk fingerprint by hash function. The client sends all chunk fingerprints to the server via the network (e.g., WAN/LAN). The server receives all fingerprints and lookups them in the index. The server returns the checking results to the client. The client sends all nonduplicate chunks to the server. The server will write them into the storage devices and update the index and metadata.

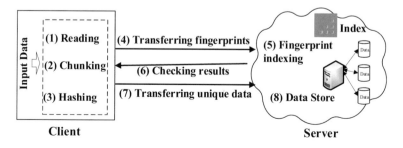

Fig. 2.8 An example of the deduplication architecture for cloud storage system

Fig. 2.9 An example of the deduplication architecture for SSD RAID

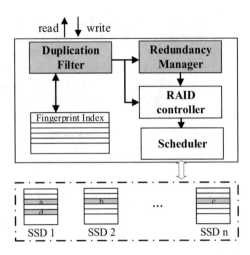

2.3.4   Deduplication for Solid State Storage

Recently, deduplication has been used in emerging nonvolatile storage devices that have endurance constraints (i.e., write limit), such as SSD devices, to reduce the amount of write traffic to the devices. CAFTL employs deduplication to eliminate duplicate I/Os to extend SSD lifetime. Nitro implements deduplication and LZ compression for SSD-cache-based primary storage to decrease the cost of SSD caches. In addition, SSD-based RAID storage improves system reliability by introducing data redundancy, which helps to tolerate cell errors as well as drive failures. Redundancy can also be used to improve read performance as creating replicas or erasure code provides more options for servicing read requests. Adopting RAID often degrades write performance as more data needs to be written to the drives.

Figure 2.9 presents an example of deduplication architecture for SSD RAID. SSD RAID consists of SSD RAID controller, fingerprint index, duplication filter, redundancy manager, scheduler, and SSDs. The RAID controller keeps performing original functionality such as striping, addressing, and redundant code computation. The duplication filter and redundancy manager are designed to enable adaptive deduplication at runtime. The duplication filter is a hardware-assisted component for detecting duplicated chunks. It maintains the fingerprints of frequently accessed data blocks in a fingerprint table. For a write request, the duplication filter compute its fingerprint and looks up it in the fingerprint index to determine if a chunk with the same content already exists in the system. Based on the current status, the redundancy manager decides if an extra redundancy copy of the chunk should be saved. If yes, the redundancy manager determines the redundancy policy and locations.

2.3.5   *Deduplication in Network Environments*

One of the initial purposes for using data deduplication is to save network bandwidth by avoiding transmitting redundant data, especially in wide area network (WAN) environments and avionics network environments. Network deduplication, also called Traffic Redundancy Elimination (TRE), is slightly different from data deduplication in storage systems. Specifically, the granularity for network deduplication, i.e., the size of a data chunk, is often tens or hundreds of bytes, which is much finer than the KB-scale (or even MB-scale) granularity in storage deduplication. Moreover, the objects for data deduplication in network environments are data streams or data packets, for which network deduplication often uses a weaker but faster hash for the fingerprinting algorithm to identify the data redundancy in a byte-by-byte manner. A study from Microsoft Research shows that packet-level redundancy elimination techniques could achieve bandwidth savings of about 15–60% on 12 large network nodes.

Figure 2.10 shows an example of the architecture of Traffic Redundancy Elimination. Replicated bytes can be detected by referring to the fingerprint cache. Then the replicated bytes are hashed to generate fingerprints and the fingerprints are transferred to server side instead of the bytes. Unlike within a storage system, Traffic Redundancy Elimination does not have the overheads of fragmented locality and garbage collection, since transferred data is typically reconstructed at the destination.

2.3.6   *Deduplication for Virtual Machines*

A great deal of redundant data exists in virtual machines, either in the main memory or the external memory. This is because the operating systems and application programs on homogeneous or heterogeneous virtual machines tend to generate duplicate data. Moreover, deduplication meets the design goal of visualization in computer systems by saving storage space of both memory and disks, thus relieving

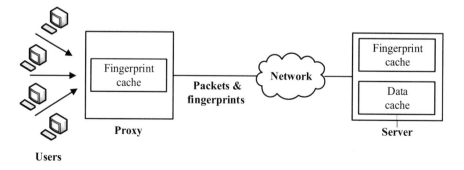

Fig. 2.10   An example of the architecture of traffic redundancy elimination

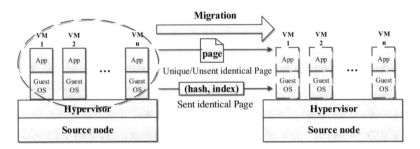

Fig. 2.11 An example of the architecture of virtual machine gang migrations via deduplication

the burden on storage devices. In addition, deduplication is employed to accelerate live migration of virtual machines in many state-of-the-art approaches by significantly reducing the amount of data migrated.

Figure 2.11 describes an example of virtual machine gang migrations via deduplication [56]. We open multiple virtual machines on top of physical machine at the source and target end. And each VM runs different workloads and application. A migration controller at the source node identifies and tracks the identical among co-located VMs. When it is time to migrate the VMs, the controller at the source node initiates the concurrent migration of all co-located VMs to the target host. Similarly, the controller at the target host prepares the host for the reception of all VMs. Furthermore, we use Super FastHash to calculate a hash for every page during the VM's memory scan. This hash value is used as a key to insert the page into the hash table.

2.4 Key Technologies of Data Deduplication

Although deduplication has been studied for more than ten years, many challenges remain. In this subsection we list the key technologies of deduplication, namely, chunking, indexing, defragmentation, security, and post-deduplication delta compression, and the challenges facing these technologies.

Chunking. Chunking algorithm has a significant impact on deduplication efficiency. Existing chunk algorithms face the challenges of high computation overheads, which renders the chunking stage the deduplication performance bottleneck and large chunk-size variance that decreases deduplication efficiency. We discuss the challenges and our proposed solutions in Chap. 3.

Indexing. Memory footprints required by fingerprint index can overflow the RAM as the size of data continues to grow. Existing indexing schemes solve the problem by leveraging either the locality or similarity. However, locality-based technologies fail to reduce the disk I/O when data stream has little or no locality, and similarity-based technologies fail to achieve high deduplication efficiency when data stream has little or no similarity. We discuss the problems and our proposed solutions in Chap. 4.

Defragmentation. Since duplicate chunks are eliminated between multiple data streams, the chunks of a stream unfortunately become physically scattered, which is known as fragmentation. Fragmentation severely decreases restore performance. Existing defragmentation schemes fail to efficiently reduce the fragmentation. We discuss the problem and our proposed technology for the problem in Chap. 5.

Security. The deduplication among data of different users can result in potential security risks. To ensure data privacy and support deduplication, users need to encrypt data and employ a secure and key management approach. Existing secure deduplication systems still face brute-force attack, heavy computation consumption on data encryption, large key space overheads, and single-point-of-failure risks of keys. We will discuss the challenges and our proposed solutions in Chap. 6.

Post-Deduplication Delta Compression. Deduplication technology can remove duplicate chunks but cannot eliminate redundancy among nonduplicate but very similar chunks. Delta compression can remove such redundancy. Thereby, delta compression is often used as a complement for data deduplication to further reduce storage or bandwidth requirements. There are challenges facing post-deduplication delta compression, including the similarity index for detecting the similar chunks can be too large to fit in RAM and high computation overheads of delta encoding. We present two schemes to address the challenges in Chap. 7.

There are many solutions in each procedure of data deduplication. In order to understand the fundamental tradeoffs in each of deduplication design choices (such as prefetching and sampling), we disassemble data deduplication into a large N-dimensional parameter space. Each point in the space is of various parameter settings and performs a tradeoff among system performance, memory footprint, and storage cost. We discuss the design tradeoff evaluation, deduplication frame-work implementation, and our design recommendation in Chap. 8.

2.5   Concluding Remarks

Due to the explosive growth of digital data in the world and the existence of large amount of redundant data in the digital data, data deduplication has become increasingly important. In this chapter we present the basic workflow of the data deduplication and its base procedure on storage systems. After that, the deduplication architecture is shown when it applies to different scenarios, including secondary storage, primary storage, cloud storage, nonvolatile storage, network environments, and virtual machines. Finally, we outline the challenges of the data deduplication. In the following chapters of this book, we will detail these challenges and present our proposed technologies to address the challenges. We will use typical real-world backup workloads to evaluate the effectiveness and efficiency of proposed technologies.

Chapter 3
Chunking Algorithms

Abstract As the first and key stage in the data deduplication workflow, chunking phase plays the role of dividing the data stream into chunks for deduplication. In this chapter, we discuss the problems facing the chunking phase and the solutions to address the problems. The rest of this chapter is organized as follows: Section 3.1 presents the state-of-the-art chunk algorithms for traffic redundancy elimination and storage systems. Section 3.2 describes the design and implementations of our proposed AE chunking algorithm for traffic redundancy elimination system. Section 3.3 describes another proposed chunking algorithm, called FastCDC, for storage system.

3.1 Existing Chunking Algorithm

Chunking algorithms can be divided into two categories: fixed-sized and content-defined. Fixed-size chunking (FSC) algorithm [17] marks chunks' boundaries by their positions and generates fixed-size chunks. This method is simple and extremely fast, but it suffers from low deduplication efficiency that stems from the boundary-shifting problem. Figure 3.1 gives an example of the problem. Data stream 2 is generated by inserting some contents at the beginning of data stream 1. If we use fixed-size chunking algorithm to process the two data streams, all chunk boundaries in data stream 2 are shifted and no duplicate chunks can be identified. We focus on content-defined chunking algorithm since it solves the boundary-shifting problem.

3.1.1 Typical Content-Defined Chunking Algorithm

Content-defined chunking (CDC) algorithm divides the data stream into variable-size chunks. It avoids the boundary-shifting problem by declaring chunk boundaries depending on local content of the data stream. If the local content is not changed, the chunks' boundaries will not be shifted. Specifically, it uses a sliding-window technique on the content of data stream and computes a hash value (e.g., Rabin

© Springer Nature Singapore Pte Ltd. 2022
D. Feng, *Data Deduplication for High Performance Storage System*,
https://doi.org/10.1007/978-981-19-0112-6_3

Fig. 3.1 An example of fixed-size chunking algorithm. Data stream 2 is generated by inserting some contents at the beginning of data stream 1

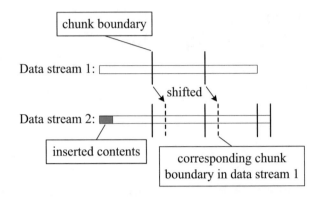

Fig. 3.2 The sliding window technique for the CDC algorithm

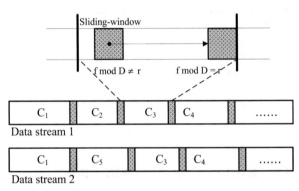

fingerprint) of the window, as shown in Fig. 3.2. The hash value of the sliding window, f, is computed via the Rabin algorithm. If the lowest log_2D bits of the hash value match a threshold value r, i.e., $f \bmod D = r$, the offset is marked as a chunk boundary (also called a cut-point). Therefore, to chunk a data stream 2 that is modified on the chunk $C2$ of data stream 1, the CDC algorithm can still identify the correct boundary of chunks $C3$ and $C4$ whose content has not been modified.

To provide the necessary basis to facilitate the discussion of and comparison among different CDC algorithms, we list below some key properties that a desirable CDC algorithm should have.

1. **Content defined.** To avoid the loss of deduplication efficiency due to the boundary-shifting problem, the algorithm should declare the chunk boundaries based on local content of the data stream.
2. **Low computational overhead.** CDC algorithms need to check almost every byte in a data stream to find the chunk boundaries. This means that the algorithm execution time is approximately proportional to the number of bytes of the data stream, which can take up significant CPU resources. Hence, in order to achieve higher deduplication throughput, the chunking algorithm should be simple and devoid of time-consuming operations.

3. **Small chunk size variance.** The variance of chunk size has a significant impact on the deduplication efficiency. The smaller the variance of the chunk size, the higher the deduplication efficiency achieved.
4. **Efficient for low-entropy strings.** The content of real data may sometimes include low-entropy strings. These strings include very few distinct characters but a large amount of repetitive bytes. In order to achieve higher deduplication efficiency, it is desirable for the algorithm to be capable of finding proper chunk boundaries in such strings and eliminating them as many as possible.
5. **Less artificial thresholds on chunk size.** Minimum and maximum thresholds are often imposed on chunk size to avoid chunks being too short or too long. These measures reduce chunk size variance, but also make the chunk boundaries position-dependent and thus not truly content-defined, which also reduces the deduplication efficiency.

Rabin-Based Chunking Algorithm. Rabin-based chunking algorithm is the most popular algorithm for computing the hash value of the sliding window in CDC for data deduplication. Specifically, a Rabin fingerprint for a sliding window (byte sequence $B_1, B_2, \ldots, B_n$ of the data stream) is defined as:

$$\mathbf{Rabin}(B_1, B_2, \ldots, B_n,) = \left\{ \sum_{x=1}^{n} B_x p^{n-x} \right\} \bmod D$$

where D is the average chunk size and n is the number of bytes in the sliding window. The Rabin fingerprint is obtained by a rolling hash algorithm since it is able to compute the fingerprint in an incremental fashion, i.e., the current hash can be incrementally computed from the previous value as follows:

$$\mathrm{Rabin}(B_1, B_2, \ldots, B_n,) = \left\{ \left[\mathrm{Rabin}(B_1, B_2, \ldots, B_n) - B_1 p^{n-1} \right] p + B_{n+1} \right\} \bmod D$$

Rabin-based chunking algorithm often imposes a minimum and a maximum threshold on chunk size to avoid the chunks being too short or too long [18]. This is because very short chunks imply more fingerprints to be stored and processed and thus not cost-effective, while very long chunks reduce the deduplication efficiency.

Rabin-based chunking algorithm is robust in finding suitable cut-points to resist the boundaries-shifting problem. However, it faces the problems of (1) low chunking throughput that renders the chunking stage the deduplication performance bottleneck and (2) large chunk-size variance that decreases deduplication efficiency.

TTTD Chunking Algorithm. Recognizing the impact of the chunk-size variance on deduplication efficiency, TTTD chunking algorithm [59], an optimized Rabin-base chunking algorithm, was proposed to improve Rabin's deduplication efficiency. To reduce the chunk-size variance, TTTD introduces an additional backup divisor that has a higher probability of finding cut-points. When it fails to find a cut-point using the main divisor within a maximum threshold, it returns the cut-point found by the backup divisor, if any. If no cut-point is found by either of the divisors, it returns the maximum threshold. Moreover, TTTD uses a larger minimum threshold and a smaller maximum threshold than the typical Rabin-based chunking algorithm, which also contributes to smaller chunk-size variance. TTTD has smaller chunk-size variance than Rabin-based chunking algorithm, while suffers from low chunking throughput.

Gear-Based Chunking Algorithm. Gear-based chunking algorithm [34] is one of the fastest chunking algorithms. Compared with Rabin-based chunking algorithm, it uses fewer operations to generate rolling hashes by means of a small random integer table to map the values of the byte contents, so as to achieve higher chunking throughput. Specifically, it employs an array of 256 random 64-bit integers to map the values of the byte contents in the sliding window and uses the addition ("+") operation to add the new byte in the sliding window into Gear hashes and the left-shift ("<<") operation to strip away the last byte of the last sliding window (e.g., B_{i-1}. More details of the algorithm are presented in 6.2.2). Gear-based chunking algorithm can attain high chunking throughput but faces the problem of low deduplication efficiency due to limited sliding window size.

MAXP Chunking Algorithm. MAXP is a state-of-the-art CDC algorithm, which is often used in remote differential compression of files and traffic redundancy elimination. It attempts to find the strict local extreme values in a fixed-size symmetric window, and then uses these points as chunk boundaries to divide the input stream. The main disadvantages of this strategy is that when declaring an extreme value, the algorithm must move backwards by a fixed-size region to check if there is any value greater (if the extreme value is the maximum value) than the value of the current position being examined. This backtracking process requires many extra conditional branch operations and increases the number of comparison operations for each byte examined. Since MAXP needs to check every byte in the input stream, any additional conditional branch operations result in a decreased chunking throughput. The advantage of MAXP algorithm is its high deduplication efficiency due to smaller chunk-size variance.

SampleByte Chunking Algorithm. EndRE proposes an adaptive Sample-Byte algorithm for redundancy elimination. The SampleByte algorithm combines the CDC algorithm's robustness to small changes in content with the efficiency of the FSC algorithm. It uses one byte to declare a fingerprint and stores $\frac{1}{p}$ representative fingerprints for content matching, where p is the sampling period. To avoid over-sampling, it skips $\frac{p}{2}$ bytes when a fingerprint has been found. SampleByte is fast since it (1) only needs one conditional branch per byte to judge the chunk boundaries and (2) skips about one-third of bytes on the input data. However, the design principle of SampleByte dictates that the sampling period p be smaller than 256 bytes, which means that the expected average chunk size must be smaller than 256 bytes when used in the chunk-level deduplication. Unfortunately, a chunk granularity of 256 bytes or smaller is too fine to be cost-efficient or practical, which makes the SampleByte algorithm inappropriate for coarse-grained chunk-level deduplication. Moreover, it requires a lookup table that needs to be generated

Table 3.1 Average throughput of the typical chunking and fingerprinting approaches in the i7-930 and i7-4770 CPUs

i7-930		i7-4770	
Rabin	SHA-1	Rabin	SHA-1
354 MB/s	526 MB/s	463 MB/s	888 MB/s

before deduplication according to the workload characteristics, which further restricts its applications.

In summary, SampleByte cannot be applied to coarse-granularity deduplication, and the other CDC algorithms suffer from low throughput problem. To make full use of the computation resources in multicore- or many core-based computer systems and accelerate the throughput of deduplication system, some deduplication systems, such as ZFS, THCAS, and DDFS, pipeline the deduplication workflow. As a result, the stage that has the lowest throughput performance will be the performance bottleneck of the whole deduplication workflow.

The throughput indexing is much higher than the chunking and fingerprinting stages when chunk size is a few KB. To analyze the performance bottleneck of the deduplication workflow, we evaluate the average throughput of the chunking and fingerprinting stages in the i7-930 and i7-4770 CPUs, with evaluation results summarized in Table 3.1, where Rabin and SHA-1 are the most common methods used in the chunking and fingerprinting stages respectively of today's deduplication storage systems. We can draw the following observations from this table. First, chunking is the main performance bottleneck of the deduplication workflow since its throughput is much lower than the fingerprinting throughput. Second, the large difference between the chunking throughput and the fingerprinting throughput means that the improvement of system throughput will be significant if the chunking-throughput performance bottleneck is removed.

The various problems facing the existing CDC algorithms motivate us to propose new chunking algorithms. We proposed Asymmetric Extremum (AE) chunking algorithm[64], a new CDC algorithm offering high throughput and small chunk-size variance thus high deduplication efficiency. AE chunking algorithm can be used for traffic redundancy elimination in network and deduplication in storage system. For storage system, we further proposed FastCDC, an extremely fast CDC algorithm that offers comparable deduplication efficiency to the typical Rabin-based chunking algorithm [66].

3.2   Asymmetric Extremum CDC Algorithm

3.2.1   The AE Algorithm Design

AE finds the local extremum value in a variable-size asymmetric window for boundaries declaring. In AE, a byte has two attributes: position and value. Each byte in the data stream has a position number, and the position of the nth byte ($1 \leq n \leq stream\ length$) in the stream is n. Each interval of S consecutive characters/bytes in the data stream is treated as a value. For example, eight consecutive characters/bytes are converted to a value that is a 64-bit integer. The value of every such interval in the data stream is associated with the position of the first byte of the S consecutive characters/bytes that constitute this value. Therefore, each byte in the stream, except for the very last $S - 1$ bytes, has a value associated with it. For convenience of description, we assume that data stream starts from the leftmost byte. If a byte A is on the left of byte B, A is said to be *before B*, and B appears *after A*. Given a byte P in the data stream, the w consecutive bytes immediately after P are defined to be the *right window* of P, and w is referred to as the *window size*.

The extreme value in the AE algorithm can be either the maximum value or the minimum value. For convenience of discussion, in what follows in this section, we assume that the extreme value is the maximum value. Starting from the very first byte of the stream or the first byte after the last cut-point (chunk boundary), AE attempts to find the first byte of the data stream that satisfies the following two conditions:

- It is first byte or its value is greater than the values of all bytes *before* it.
- Its value is *not less than* the values of all bytes in its *right window*.

The first byte found to meet these conditions is referred to as a *maximum point*. These two conditions make sure that the maximum point has the maximum value in the region from the very first byte of the stream or the first byte after the last cut-point to the rightmost byte of the maximum point's *right window*. There are two further implications. First, this first byte can be a maximum point. Second, AE allows for ties between the byte being examined and bytes in its *right window*. If a maximum point has been found, AE returns the rightmost byte in its *right window*, which is also the byte being processed, as a cut-point (chunk boundary). AE does not need to backtrack, since the process after the returned cut-point is independent of the content before the cut-point. Moreover, the position and the value of the bytes processed (except for the byte having the temporary local maximum value) need not be kept in memory. The workflow of AE is described in Fig. 3.3. Note that N, M, D in the figure are neighboring positions, and so are the positions of C and B$'$.

Algorithm 1 below provides a more detailed implementation of the AE chunking algorithm. The minimum chunk size and the expected chunk size of AE are respectively $w + 1$ and $(e - 1)\ w$.

Fig. 3.3 The workflow of the AE chunking algorithm

Step 1: Input data stream

Start from B, B is the first byte of the input data stream

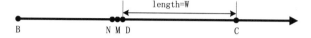

B

Step 2: Searching for maximum point

M is extreme point if:
1) The interval [B, N] is empty, or the value of M is greater than the values of all bytes in the interval [B, N].
2) The value of M is no less than the values of all bytes in the interval [D, C]

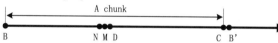

length=W

B   N M D   C

Step 3: Declaring chunk boundary

Return C as cut-point. B' is the first byte of the remaining input stream

A chunk

B   N M D   C B'

Algorithm 1: Algorithm of AE Chunking

Input: input string, Str; left length of the input string, L;
Output: chunked position (cut-point), i;
 1: Predefined values: window size w;
 2: **function** AECHUNKING(Str, L)
 3:     $i \leftarrow 1$
 4:     $max.value \leftarrow Str[i].value$
 5:     $max.position \leftarrow i$
 6:     $i \leftarrow i + 1$
 7:     **while** $i < L$ **do**
 8:         **if** $Str[i].value \leq max.value$ **then**
 9:             **if** $i = max.position + w$ **then**
10:                 **return** i
11:             **end if**
12:         **else**
13:             $max.value \leftarrow Str[i].value$
14:             $max.position \leftarrow i$
15:         **end if**
16:         $i \leftarrow i + 1$
17:     **end while**
18:     **return** L
19: **end function**

3.2.2   The Optimized AE Algorithm

In fact, AE uses the comparison operation to find the extreme point, which provides us with an opportunity to optimize the algorithm. As shown in Algorithm 1, every time the algorithm moves forward one byte, the value of the new byte $Str[i].value$ must be compared with the temporary maximum value $max.value$. However, $max.value$, which is selected out of the values of all bytes between the starting point and the current position i, is expected to be greater than $Str[i].value$. In other words, lines 8 through 10 in the Algorithm 1 are expected to be executed much more frequently than lines 12 through 14, which means that optimizing lines 8 through 10 in the algorithm will significantly improve the chunking throughput.

We found that declaring boundary by Algorithm 1 (line 9) is unnecessary in the region between $max.position$ and $max.position + w$ due to AE's strategy of declaring boundaries. Therefore, we optimize the algorithm by removing as many of the boundary declaring operations as possible in the region mentioned above as described in Algorithm 2. For simplicity, we use $max.v$ and $max.pos$ instead of $max.value$ and $max.position$ in Algorithm 1. This optimization helps increase the chunking throughput since it only needs one comparison and one conditional branch in the region between $max.position$ and $max.position + w$ when $max.value$ is greater than $Str[i].value$, rather than one comparison and two conditional branches required by the original AE algorithm.

Algorithm 2: The Optimized AE

Input: input string, Str; left length of the input string, L;
Output: chunked position (cut-point), i;
1: Predefined values: window size w;
2: **function** AECHUNKING-OPT(Str, L)
3:      $i \leftarrow 1$
4:      $max.v \leftarrow Str[i].v$
5:      $max.pos \leftarrow i$
6:      $i \leftarrow i + 1$
7:      **while** $i < L$ **do**
8:          **if** $L < max.pos + w$ **then**
9:              $endPos = L$
10:         **else**
11:             $endPos = max.pos + w$
12:         **end if**
13:         **while** $Str[i].v < max.v$ and $i < endPos$ **do**
14:             $i \leftarrow i + 1$
15:         **end while**
16:         **if** $Str[i].v > max.v$ **then**
17:             $max.v \leftarrow Str[i].v$
18:             $max.pos \leftarrow i$
19:         **else**
20:             **return** $max.pos + w$
21:         **end if**
22:         $i \leftarrow i + 1$
23:     **end while**
24:     **return** L
25: **end function**

3.2.3 Properties of the AE Algorithm

We analyze the AE algorithm in regards to the key properties of CDC algorithms.

Content Defined. The MAXP algorithm considers a byte with the local maximum value a chunk boundary. Therefore, any modifications within a chunk, as long as they do not replace the local maximum value, will not affect the adjacent chunks, since the chunk boundaries will simply be realigned. Unlike MAXP, the AE algorithm returns the Wth position after the maximum point as the chunk boundary. It puts the maximum points inside the chunks instead of considering them as chunk boundaries. This strategy may slightly decrease the deduplication efficiency, but AE is still content-defined since the maximum points inside the chunks can also realign the chunk boundaries.

Fig. 3.4 An example of
efficiency loss of AE

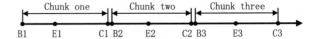

Table 3.2 Computational overheads of the three algorithms (where p is the expected chunk size)

Algorithm	Computational overhead per byte scanned
Rabin	1 or, 2 xors, 2 shifts, 2 array lookups, 1 conditional branch
MAXP	2 mod, $2\frac{1}{p}$ comparisons, $5+\frac{1}{p}$ conditional branches
AE	1 comparison, 2 conditional branches

Fig. 3.5 Illustration of the
key difference between the
MAXP and AE algorithms.
(a) MAXP. **(b)** AE

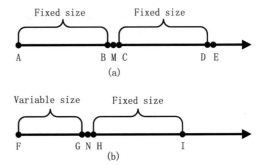

Take Fig. 3.4 for example, E1, E2, E3 are the three maximum points and C1, C2, C3 are the cut-points of the three corresponding chunks. Assume that all modifications in the example will not replace the local maximum value. If there is an insertion (or deletion) in the interval [B1, E1) in Chunk 1, Chunk 2 will not be affected since the chunk boundary will be realigned by the maximum point E1. If the insertion is in the interval (E1, C1], the starting point of Chunk 2 will be changed, and E2 will realign the boundary to keep Chunk 3 from being affected. If a sequence of consecutive chunks has been modified, the loss of efficiency is determined by the position of the modification in the last modified chunk. If the modification is *before* the maximum point, there is no efficiency loss. Otherwise, only one duplicate chunk that is immediately after this modified region will be affected. In addition, the deduplication efficiency is also determined by many other factors, such as chunk-size variance and the ability to eliminate low-entropy strings. AE's ability to eliminate low-entropy strings and reduce chunk-size variance has more than compensated for this relatively small loss of deduplication efficiency.

Computational Overheads. Table 3.2 shows the computational overheads of the three algorithms, AE, MAXP, and Rabin. As shown in the table, the Rabin algorithm needs 1 OR, 2 XORs, 2 SHIFTs, and 2 ARRAY LOOKUPs per byte examined to compute the fingerprints and one conditional branch to judge the chunk boundaries. While both the MAXP and AE algorithms use comparison operations to find the local maximum values, their strategies are quite different and it is this difference that makes AE much faster than MAXP. Figure 3.5 shows the difference between the two algorithms. B, M, C in the figure are neighboring positions, so are the positions of D and E and positions of G, N, and H. As shown in Fig. 3.5, MAXP finds the maximum values in a fixed-size window [A, D]. If the byte M that is in the center of this window has the maximum value in the window, its value must be strictly greater than that of any byte in both regions of [A, B] and [C, D]. Assuming that all bytes in the window [A, D] have been scanned and M has the maximum value and has been returned as a cut-point, some of the bytes in region [C, D] must be scanned again when MAXP processes the byte E. This means that MAXP needs an array to store the information of the bytes in the fixed-size region immediately before the current byte. Therefore, it requires two modular operations to update the array, and $2 - \frac{1}{p}$ comparison and $5 + \frac{1}{p}$ conditional branch operations to find the local maximum value.

In contrast, AE only needs to find the maximum value in an asymmetric window [F, I], which includes a fixed-size region [H, I] and a variable-size region [F, G], whose size is determined by the content of the data stream. As a result, we only need to store a candidate maximum point and the position of the candidate maximum point, and do not need to backtrack to declare the local maximum value. Therefore, AE only needs one comparison and two conditional branch operations. Clearly, AE requires much fewer operations, particularly the time-consuming conditional branch and table lookup operations, than the other two algorithms.

Chunk-Size Variance. Next we analyze the chunk size variance of the AE algorithm. We use the probability of a long region not having any cut-point to estimate the chunk size variance.

Theorem 1 *AE has no maximum point in a given range, if and only if in each interval of w consecutive bytes in this range, there exists at least one byte that satisfies the first condition of the maximum point, namely, it is the first byte or its value is greater than the values of all bytes before it.*

Proof In this range, if there exists one byte in each interval of w consecutive bytes whose value is greater than the values of all bytes before it, then the second condition of the maximum point, namely, its value is *not less than* the values of all bytes in its right window, will never be satisfied. In other words, there is no maximum point in the range.

Given an interval $[cw + a + 1, cw + a + w]$, where c is a constant, the probability of no byte satisfying the first condition of maximum point is:

$$\prod_{i=1}^{w} \left(1 - \frac{1}{a + i}\right) = \frac{a}{a + w}$$

So the complementary probability, that there exists at least one byte satisfying the first condition of the maximum point, is $\frac{w}{w+a}$. Divide the interval into subintervals with the length of w and then number them from 1 to m. Consider the subinterval $[(p - 1)w + 1, pw]$. The probability of no maximum point in it is:

Table 3.3 Probability of no cut-points in a region of length $m \times$ *average-chunk-size*

m	AE	Rabin_0	Rabin_0.25	MAXP
	$\frac{1}{[(e-1)*m]!}$	e^{-m}	$e^{-1.2m}$	$\frac{2^{2m}}{(2m)!}$
2	0.0938	0.1353	0.0907	0.6667
3	0.0064	0.0498	0.0273	0.0889
4	2.56×10^{-4}	0.0183	0.0082	0.0063
5	6.85×10^{-6}	0.0067	0.0025	2.82×10^{-4}
6	1.32×10^{-4}	0.0025	7.47×10^{-4}	8.55×10^{-6}
7	1.94×10^{-9}	9.12×10^{-4}	2.25×10^{-4}	1.88×10^{-7}
8	2.25×10^{-11}	3.35×10^{-4}	6.77×10^{-5}	3.13×10^{-9}

$$\frac{w}{(p-1) \times w + w} = \frac{1}{p}$$

Multiplying the probabilities of the continuous m subintervals, we have $\frac{1}{m!}$. Given that the average chunk size of AE is $(e-1)w$, the probability of no maximum point in m consecutive chunks of average chunk size becomes:

$$p(\text{AE}) = \frac{1}{[(e-1) \times m]!}$$

here m should be more than 1.

Next we compare the probabilities of very long chunks among the AE, MAXP and Rabin algorithms. Table 3.3 shows formulas to calculate the theoretical probability of no cut-points in a region of length $m \times average - chunks - size$ and lists such probabilities when $m = 2, 3, \cdots, 8$ for the three algorithms, where Rabin 0 represents the Rabin algorithm without minimum threshold, and Rabin_0.25 represents Rabin with a minimum threshold on chunk size, and the ratio of the minimum threshold to the expected chunk size is 0.25. As can be seen from the table, the probability of generating exceptionally long chunks by AE is much lower than the other two algorithms, which also means that AE has smaller chunk-size variance.

Dealing with Low-Entropy Strings. Ties between the byte being examined and the bytes in the *right window* may appear in the data stream. If a tie happens to be between two local maximum values, we can break the tie by one of the following two strategies: (1) selecting the first maximum value or (2) going beyond the right window to search for a strictly maximum value. Strategy (1) can help identify and eliminate low-entropy strings. AE allows for ties in its right window and the maximum point can be the first byte, so that it can divide the low-entropy strings into fixed-size chunks whose size is $w + 1$. On the other hand, Strategy (2), which is used in the MAXP algorithm, will lead the algorithm to miss detecting and eliminating low-entropy strings. Note that the AE algorithm cannot detect all low-entropy strings. If the length of a low-entropy string is greater than $2w + 2$, then AE can identify a part of it. Furthermore, Strategy (2) requires more conditional branch operations in finding the maximum points. For these reasons, we chose Strategy (1) for AE.

3.2.4 Performance Evaluation

We present the experimental evaluation of our AE algorithm in terms of multiple performance metrics. To characterize the benefits of the AE and optimized AE algorithm, we also compare it with two state-of-the-art CDC algorithms, namely, Rabin and MAXP.

Evaluation Setup. We impose a minimum threshold and a maximum threshold on Rabin's chunk size whose values are $\frac{1}{4} \times$ and $8\times$ the expected average chunk size respectively. We represent the Rabin algorithm with such configurations by Rabin 0.25. Because of the minimum threshold, the real average chunk size of Rabin will be greater than the expected average chunk size. It is approximately equal to the expected average chunk size plus the minimum threshold. For the sake of fairness, for each test, we first processed using Rabin to get the real average chunk size, and then adjusted the real average chunk size to the same value when using other algorithms. For convenience of discussion, Rabin's expected average chunk sizes are used as labels to distinguish different tests on each dataset.

Datasets. To evaluate the three CDC algorithms, we use the following four real-world datasets.

1. *Network traffic dataset:* The network traffic in this dataset was collected using wireshark from our research laboratory. The trace collection spanned a period of 7 days and yielded about 53G of data.
2. *Similar movie-files dataset:* This dataset is composed of 24 movie files constituting 12 different movies with different subtitle or dubbing. The total size of this dataset is 31G.
3. *TAR dataset:* This dataset includes 20 versions of GCC, 35 versions of GLIB, 15 versions of GDB, 10 versions of EMACS, and 40 versions of Linux kernels. Each version of these free software was packaged as a tar file. The total size of this dataset is 32G.
4. *VMDK dataset:* This dataset consists of 1.85T of 125 backups of an Ubuntu virtual machine. Since all backups are full backup, there exists a large amount of duplicate content in this dataset.

Deduplication Efficiency

We use deduplication elimination ratio (DER), which we define as the ratio of the size of input data to the size of data need to be actually stored/transferred, to measure the deduplication efficiency. Therefore, the greater the value of DER, the higher the deduplication efficiency is.

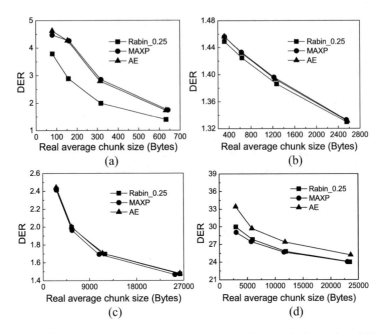

Fig. 3.6 Deduplication efficiency of the three chunking algorithms on the four real-world datasets. (**a**) Traffic. (**b**) Movies. (**c**) Tar. (**d**) VMDK

Figure 3.6 shows the results of this comparison. In the figure we can see that AE achieves comparable deduplication efficiency to MAXP on the first two datasets, and both AE and MAXP outperform Rabin in the DER measure. On the third dataset (Fig. 3.6c), AE achieves almost the same efficiency as Rabin, and their deduplication efficiency are slightly higher than MAXP. On the last data set (Fig. 3.6d), AE achieves higher DER than the other two algorithms. The main reason for AE's superior DER performance is its ability to detect and eliminate more low-entropy strings and smaller chunk size variance.

Chunking Throughput

Figure 3.7 compares the chunking throughput between the Rabin, MAXP, AE, and the optimized AE algorithms on the four datasets. The expected average chunk sizes used on the four datasets are 2 KB, 4 KB, 8 KB, and 16 KB respectively, which serves to test the sensitivity of the algorithms to chunk size.

The result shows that the AE algorithm outperforms the other two algorithms in terms of the chunking throughput, and the optimized AE algorithm significantly improves the chunking throughput of the AE algorithm. Specifically, AE achieves a throughput of more than 1.4 GB/s, which is about 3.2 times higher than Rabin_0.25, and 2.3 times higher than MAXP. The optimized AE algorithm further accelerates the throughput of AE by about 1.75 times. Moreover, Fig. 3.6 also indicates that the

Fig. 3.7 Chunking throughput of the four algorithms on the four datasets

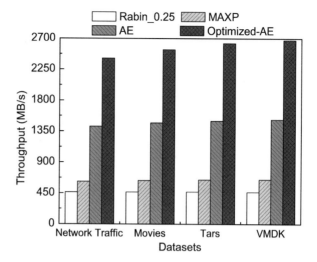

chunking throughput of the AE algorithm improves as the average chunk size increases.

3.3   FastCDC: A Fast and Efficient CDC Approach

3.3.1   Limitation of Gear-Based Chunking Algorithm

Gear-based CDC is first employed by Ddelta [34] for delta compression, which helps provide a higher delta encoding speed. However, according to our experimental analysis, there are still challenges facing the Gear-based CDC. We elaborate on these issues as follows.

Limited Sliding Window Size. The traditional hash judgment for the Rabin-based CDC is also used by the Gear-based CDC algorithm. But this results in a smaller-sized sliding window used by Gear-based CDC since it uses Gear hash for chunking. For example, as shown in Fig. 3.8, the sliding window size of the Gear-based CDC will be equal to the number of bits used by the mask value. Therefore, when using a mask value of 213 for the expected chunk size of 8 KB, the sliding window for the Gear-based CDC would be 13 bytes, while that of the Rabin-based CDC would be 48 bytes [18]. The smaller sliding window size of the Gear-based CDC can lead to more chunking position collisions (i.e., randomly marking the different positions as the chunk cut-points), resulting in the decrease in deduplication ratio.

The Time-Consuming Hash Judgment. Our implementation and in-depth analysis of the Gear-based CDC suggest that its hash-judging stage accounts for more than 60% of its CPU overhead during CDC after the fast Gear hash is used for fingerprinting. Thus, there is a lot of room for the optimization of the hash judging stage to further accelerate the CDC process.

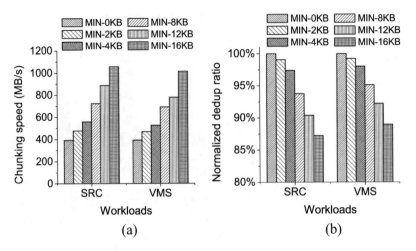

Fig. 3.8 Rabin-based CDC performance as a function of the minimum chunk size used for cut-points skipping before chunking. (**a**) Chunking speed. (**b**) Deduplication ratio

Table 3.4 Workload characteristics of the seven datasets

Name	Size	Workload description
TAR	19 GB	85 tarred files from the open source projects such as GCC, GDB, Emacs.
LNX	105 GB	260 versions of Linux source code files.
WEB	36 GB	15 days' snapshots of the website: news.sina.com, which are collected by crawling software wget with a maximum retrieval depth of 3.
VMA	117 GB	75 virtual machine images of different OS release versions, including CentOS, Fedora, Debian.
VMB	1.9 TB	125 backups of an Ubuntu 12.04 virtual machine image in use by a research group.
RDB	1.1 TB	200 backups of the redis key-value store database.
SYN	1.4 TB	200 synthetic backups. The backup is simulated by the file create/delete/modify operations.

Speed Up Chunking by Skipping. Another observation is that the minimum chunk size used for avoiding extremely small-sized chunks, can be also employed to speed up CDC by the cut-point skipping, i.e., eliminating the chunking computation in the skipped region. Figure 3.8 shows our experimental observation of Rabin-based CDC with two typical workloads of deduplication whose workload characteristics are detailed in Table 3.4 in Sect. 3.3.7. The evaluation uses the average chunk size of 8 KB, Intel i7-4770 processor, and the best open-source Rabin algorithm we have access to for the speed test. Figure 3.8a indicates that setting the minimum chunk size for cut-point skipping at $\frac{1}{4} \times -2 \times$ of the expected chunk size can effectively accelerate the CDC process. But this approach decreases the deduplication ratio by about 2–15% (see Fig. 3.8b) since many chunks are not divided truly according to the data contents, i.e., not really content-defined.

The observation suggested in Fig. 3.8 motivates us to consider a new CDC approach that (1) keeps all the chunk cut-points that generate chunks larger than a predefined minimum chunk size and (2) enables the chunk size distribution to be normalized to a relatively small specified region, an approach we refer to as normalized chunking.

3.3.2   FastCDC Overview

FastCDC is implemented on top of the Gear-based CDC, and aims to provide high performance CDC. Generally, there are three metrics for evaluating CDC performance, namely, deduplication ratio, chunking speed, and the average generated chunk size.

Note that the average generated chunk size may be nearly equal to or larger than the predefined expected chunk size (e.g., 8 KB) due to factors such as the detailed CDC methods and datasets. This is also an important CDC performance metric because it reflects the metadata overhead for deduplication indexing, i.e., the larger the generated chunk size is, the fewer the number of chunks and thus the less metadata will be processed by data deduplication. However, it is difficult, if not impossible, to improve these three performance metrics simultaneously because they can be conflicting goals. For example, a smaller average generated chunk size leads to a higher deduplication ratio, but at the cost of lower chunking speed and high metadata overheads. Thus, FastCDC is designed to strike a sensible tradeoff among these three metrics so as to strive for high-performance CDC, by using a combination of the three techniques with their complementary features as shown in Fig. 3.9.

- Optimizing hash judgment: using a zero-padding scheme and a simplified hash-judging statement to speed up CDC without compromising the deduplication ratio, as detailed in Sect. 3.3.3.
- Sub-minimum chunk cut-point skipping: enlarging the predefined minimum chunk size and skipping cut-points for chunks smaller than that to provide a

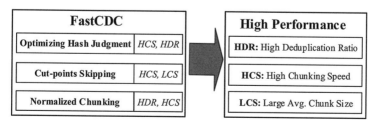

Fig. 3.9 The three key techniques used in FastCDC and their corresponding benefits for high-performance CDC

higher chunking speed and a larger average generated chunk size, as detailed in
Sect. 3.3.4.
- Normalized chunking: selectively changing the number of mask '1' bits for the
  hash judgment to approximately normalize the chunk-size distribution to a small
  specified region that is just larger than the predefined minimum chunk size,
  ensuring both a higher deduplication ratio and higher chunking speed, as detailed
  in Sect. 3.3.5.

In general, the key idea behind FastCDC is the combined use of the above three
key techniques on top of Gear-based CDC, especially employing normalized
chunking to address the problem of decreased deduplication ratio facing the
cut-point skipping, and thus achieve high-performance CDC on the three key metrics

3.3.3   Optimizing Hash Judgment

In this subsection, we propose an enhanced but simplified hash-judging statement to
accelerate the hash judgment stage of FastCDC to further accelerate the chunking
process on top of the Gear-based CDC and increase the deduplication ratio to reach
that of the Rabin-based CDC. More specifically, FastCDC incorporates two main
optimizations as elaborated below.

Enlarging the Sliding Window Size by Zero Padding. The Gear-based CDC
employs the same conventional hash judgment used in the Rabin-based CDC,
where a certain number of the lowest bits of the fingerprint are used to declare the
chunk cut-point, leading to a shortened sliding window for the Gear-based CDC
because of the unique feature of the Gear hash, as shown in Fig. 3.10. In the figure,
because of the computation principles of the Gear hash, the size of the sliding
window used for hash judgment is only 5 bytes. To address this problem, FastCDC
enlarges the sliding window size by padding a number of zero bits into the mask
value.

As illustrated by the example of Fig. 3.11, FastCDC pads 5 zero bits into the mask
value and changes the hash judgment statement to "fp & $mask == r$". If the masked

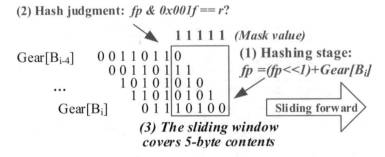

Fig. 3.10   An example of the sliding window technique used in the Gear-based CDC

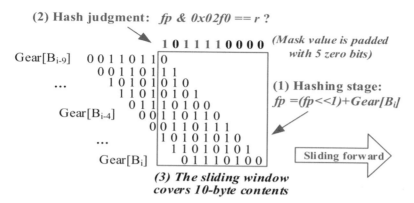

Fig. 3.11 An example of the sliding window technique proposed for FastCDC

bits of fp match a threshold value r, the current position will be declared as a chunk cut-point. Since Gear hash uses one left-shift and one addition operation to compute the rolling hash, this zero-padding scheme enables 10 bytes (i.e., $B_i . . . B_{i+9}$), instead of the original 5 bytes, to be involved in the final hash judgment by the five masked 1 bits (as the red box shown in Fig. 3.11) thus making the sliding window size equal or similar to that of the Rabin-based CDC [18], minimizing the probability of the chunking position collision. As a result, FastCDC is able to achieve a deduplication ratio as high as that by the Rabin-based CDC.

Simplifying the Hash Judgment to Accelerate CDC. The conventional hash judgment process, as used in the Rabin-based CDC, is expressed in the programming statement of "*fp mod D == r*" [18, 34]. For example, the Rabin-based CDC usually defines D and r as $0x02000$ and $0x78$, according to the known open source project LBFS [18], to obtain the expected average chunk size of 8 KB. In FastCDC, when combined with the zero padding scheme introduced above and shown in Fig. 3.11, the hash judgment statement can be optimized to "*fp & Mask == 0*", which is equivalent to "*!fp & Mask*". Therefore, FastCDC's hash judgment statement reduces the register space for storing the threshold value r and avoids the unnecessary comparison operation that compares "*fp & Mask*" and r, thus further speeds up the CDC process.

3.3.4 Cut-Point Skipping

Most of the CDC-based deduplication systems impose a limit of the maximum and minimum chunk sizes, to avoid the pathological cases of generating many extremely large- or small-sized chunks by CDC [4, 18, 67–70]. A common configuration of the average, minimum, and maximum parameters follows that used by LBFS [18], i.e., 8 KB, 2 KB, and 64 KB. Our experimental observation and mathematical analysis suggest that the cumulative distribution of chunk size X in Rabin-based CDC

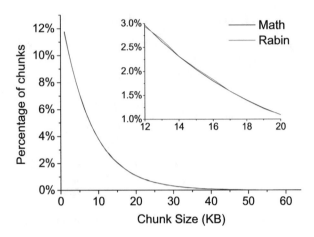

Fig. 3.12 Chunk-size distribution of the Rabin-based CDC approach with average chunk size of 8 KB and without the maximum and minimum chunk size requirements

approaches with an expected chunk size of 8 KB (without the maximum and minimum chunk size requirements) follows an exponential distribution as follows:

$$P(X \le x) = F(x) = \left(1 - e^{-\frac{x}{8192}}\right), x \ge 0$$

Note that this theoretical exponential distribution assumes that the data content and Rabin hashes of contents follow a uniform distribution. This equation suggests that the value of the expected chunk size will be 8 KB according to the exponential distribution.

Figure 3.12 shows a comparison between the actual chunk-size distribution of the real-world datasets after the Rabin-based CDC and the chunk-size distribution obtained by the mathematical analysis, which indicates that the two are almost identical. The chunks smaller than 2 KB and larger than 64 KB would account for about 22.12% and 0.03% of the total number of chunks respectively. This means that imposing the maximum chunk size requirement only slightly hurts the deduplication ratio but skipping cut-points before chunking to avoid generating chunks smaller than the prescribed minimum chunk size, or called sub-minimum chunk cut-point skipping, will impact the deduplication ratio significantly as evidenced in Fig. 3.12. This is because a significant portion of the chunks are not divided truly according to the data contents, but forced by this cut-point skipping.

Given FastCDC's goal of maximizing the chunking speed, enlarging the minimum chunk size and skipping sub-minimum chunk cut-point will help FastCDC achieve a higher CDC speed by avoiding the operations for the hash calculation and judgment in the skipped region. This gain in speed, however, comes at the cost of reduced deduplication ratio. To address this problem, we will develop a normalized chunking approach, to be introduced in the next subsection.

It is worth noting that this cut-point skipping approach, by avoiding generating chunks smaller than the minimum chunk size, also helps increase the average generated chunk size. In fact, the average generated chunk size exceeds the expected chunk size by an amount equal to the minimum chunk size.

3.3.5 Normalized Chunking

In this subsection, we propose a novel chunking approach, called normalized chunking, to solve the problem of decreased deduplication ratio facing the cut-point skipping approach. As shown in Fig. 3.13, normalized chunking generates chunks whose sizes are normalized to a specified region centered at the expected chunk size. The dotted line shows a higher level of normalized chunking. After normalized chunking, there are almost no chunks of size smaller than the minimum chunk size, which means that normalized chunking enables skipping cut-points for sub-minimum chunks to reduce the unnecessary chunking computation and thus speed up CDC.

In our implementation of normalized chunking, we selectively change the number of effective mask bits (i.e., the number of '1' bits) for the hash-judging statement. For the traditional CDC approach with expected chunk size of 8 KB (i.e., 2^{13}), 13 effective mask bits are used for hash judgment (e.g., fp & $0x1fff == r$). For normalized chunking, more than 13 effective mask bits are used for hash judgment (e.g., fp & $0x7fff == r$) when the current chunking position is smaller than 8 KB, which makes it harder to generate chunks of size smaller than 8 KB.

On the other hand, fewer than 13 effective mask bits are used for hash judgment (e.g., fp & $0x0fff == r$) when the current chunking position is larger than 8 KB, which makes it easier to generate chunks of size larger than 8 KB. Therefore, by changing the number of '1' bits in FastCDC, the chunk-size distribution will be approximately normalized to a specified region always larger than the minimum chunk size, instead of following the exponential distribution (see Fig. 3.12).

Generally, there are three benefits or features of normalized chunking (NC):

- NC reduces the number of small-sized chunks, which makes it possible to combine it with the cut-point skipping approach to achieve high chunking speed without sacrificing the deduplication ratio as suggested in Fig. 3.12.

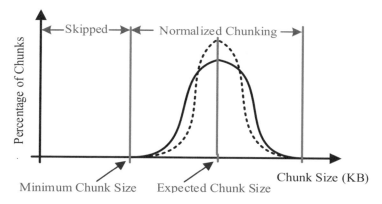

Fig. 3.13 A conceptual diagram of the normalized chunking combined with the sub-minimum chunk cut-point skipping

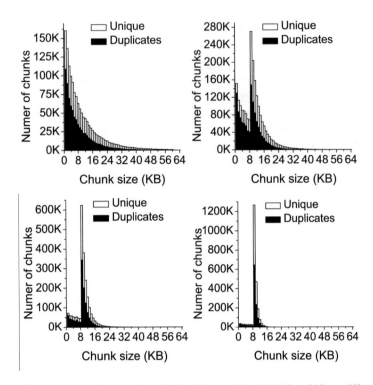

Fig. 3.14 Chunk-size distribution of FastCDC with normalized chunking (NC) at different normalization levels

- NC further improves the deduplication ratio by reducing the number of large-sized chunks, which compensates for the reduced deduplication ratio caused by reducing the number of small-sized chunks in FastCDC.
- The implementation of FastCDC does not add additional computing and comparing operations. It simply separates the hash judgment into two parts, before and after the expected chunk size.

Figure 3.14 shows the chunk-size distribution after normalized chunking in comparison with FastCDC without NC on the TAR dataset. The normalization levels 1, 2, 3 indicate that the normalized chunking uses the mask bits of (14, 12), (15, 11), (16, 10), respectively, where the first and the second integers in the parentheses indicate the numbers of effective mask bits used in the hash judgment before and after the expected chunk size (or normalized chunk size) of 8 KB. Figure 3.14 suggests that the chunk-size distribution is a reasonably close approximation of the normal distribution centered on 8 KB at the normalization level of 2 or 3.

As shown in Fig. 3.14, there are only a very small number of chunks smaller than 2 KB or 4 KB after normalized chunking while FastCDC without NC has a large number of chunks smaller than 2 KB or 4 KB. Thus, when combining NC with the cut-point skipping to speed up the CDC process, only a very small portion of chunk cut-points will be skipped in FastCDC, leading to nearly the same deduplication ratio as the conventional CDC approaches without the minimum chunk size require-ment. In addition, normalized chunking allows us to enlarge the minimum chunk size to maximize the chunking speed without sacrificing the deduplication ratio.

It is worth noting that the chunk-size distribution shown in Fig. 3.14 is not truly normal distribution but an approximation of it. Figure 3.14c, d shows a closer approximation of normal distribution of chunk size achieved by using the normal-ization levels 2 and 3. Interestingly, the highest normalization level of NC would be equivalent to Fixed-Size Chunking (FSC), i.e., all the chunk sizes are normalized to be equal to the expected chunk size. Since FSC has a very low deduplication ratio but extremely high chunking speed, it means that there will be a "sweet spot" among the normalization level, deduplication ratio, and chunking speed

3.3.6   The FastCDC Algorithm Design

To put things together and in perspective. Algorithm 3 describes FastCDC combin-ing the three key techniques: optimizing hash judgment, cut-point skipping, and normalized chunking (with the expected chunk size of 8 KB). The data structure "Gear" is a predefined array of 256 random 64-bit integers with one-to-one mapping to the values of byte contents for chunking [34].

Algorithm 3: FastCDC8KB

Require: data buffer, src; buffer length, n
Ensure: chunking breakpoint, i;

1: $MaskS \leftarrow 0x0003590703530000LL$;
2: $MaskA \leftarrow 0x0000d90303530000LL$;
3: $MaskL \leftarrow 0x0000d90003530000LL$;
4: $MinSize \leftarrow 2KB$; $MaxSize \leftarrow 64KB$;
5: $fp \leftarrow 0$; $i \leftarrow MinSize$; $NormalSize \leftarrow 8KB$
6: **if** $n \le MinSize$ **then**
7:     **return** n
8: **end if**
9: **if** $n \ge MaxSize$ **then**
10:    $n \leftarrow MaxSize$;
11: **else if** $n \le NormalSize$ **then**
12:    $NormalSize \leftarrow n$;
13: **end if**
14: **for** ; $i \le NormalSize$; $i++$ **do**
15:    $fp = (fp << 1) + Gear[src[i]]$;
16:    **if** $!(fp\&MaskS)$ **then**
17:       **return** i;
18:    **end if**
19: **end for**
20: **for** ; $i \le n$; $i++$ **do**
21:    $fp = (fp << 1) + Gear[src[i]]$;
22:    **if** $!(fp\&MaskL)$ **then**
23:       **return** i;
24:    **end if**
25: **end for**
26: **return** i;

As shown in Algorithm 3, FastCDC uses normalized chunking to divide the chunking judgment into two loops with the optimized hash judgment. Note that FastCDC without normalized chunking is not shown here but can be easily implemented by using the new hash-judging statement "!*fp* & *MaskA*", where the *MaskA* is padded with 35 zero bits to enlarge the sliding window size to 48 bytes as that used in the Rabin-based CDC [18]. Note that *MaskA*, *MaskS*, and *MaskL* are three empirically derived values, where the padded zero bits are almost evenly distributed for slightly higher deduplication ratio according to our large scale tests.

FastCDC implements normalized chunking by using mask value *MaskS* and *MaskL* to make the chunking judgment harder or easier (to generate chunks smaller or larger than the expected chunk size) when the current position is smaller or larger than the expected chunk size, respectively. And the number of '1' bits in *MaskS* and *MaskL* can be changed for different normalization levels. The minimum chunk size used in Algorithm 3 is 2 KB, which can be enlarged to 4 KB or 8 KB to further speed up the CDC process while combining with normalized chunking.

3.3.7 Performance Evaluation

Experimental Platform. To evaluate FastCDC, we implement a prototype of the data deduplication system on the Ubuntu 12.04.2 operating system running on a quadcore Intel i7-4770 processor at 3.4 GHz, with a 16GB RAM. To better evaluate the chunking speed, another quad-core Intel i7-930 processor at 2.8 GHz is also used for comparison.

Configurations for CDC and Deduplication. Three CDC approaches, Rabin-, and Gear-based CDC, are used as the baselines for evaluating FastCDC. Rabin-based CDC is implemented based on the open-source project LBFS [18] (also used in many published studies [14, 22] or project [6]), where the sliding window size is configured to be 48 bytes. The Gear-based schemes are implemented according to the algorithms described in their papers [34], and we obtain performance results similar to and consistent with those reported in these papers. Here all the CDC approaches are configured with the maximum and minimum chunk sizes of $8\times$ and $\frac{1}{4}\times$ of the expected chunk size, the same as configured in LBFS [18]. The deduplication prototype consists of approximately 3000 lines of C code, which is compiled by GCC 4.7.3 with the "-O3" compiler option.

Performance Metrics of Interest. Chunking speed is measured by the in-memory processing speed of the evaluated CDC approaches and obtained by the average speed of five runs. Deduplication ratio is measured in terms of the percentage duplicates detected after CDC, i.e., $\frac{\text{the size of duplicated data detected}}{\text{total data size before deduplication}}$. Average chunk size is $\frac{\text{total data size}}{\text{number of chunks}}$ after CDC, which reflects the metadata overhead for deduplication indexing.

Evaluated Datasets. Seven datasets with a total size of about 5TB are used for evaluation as shown in Table 3.4. These datasets consist of the various typical workloads of deduplication, including the source code files, virtual machine images, and database snapshots, whose deduplication ratios vary from 40% to 97%.

Evaluation of FastCDC. Next we evaluate the performance of FastCDC with the combined capability of the three key techniques: optimizing hash judgment, cut-point skipping, and normalized chunking. Four approaches are tested for evaluation: RC with Min2KB (or RC-MIN-2KB) is Rabin-based CDC used in LBFS; FC with Min2KB (or FC-MIN-2KB) uses the techniques of optimizing hash judgment and cut-point skipping with a minimum chunk size of 2 KB; FC-NC with Min4KB and FCNC with Min8KB refer to FastCDC using all the three techniques with a minimum chunk size of 4 KB and 8 KB, respectively. To better evaluate the deduplication ratio, Fixed-Size Chunking (XC) is also tested using the average chunk size of 10 KB.

Evaluation results in Table 3.4 suggest that FC with Min2KB achieves nearly the same deduplication ratio as Rabin-based approach. FC-NC with Min4KB achieves the highest deduplication ratio among the five approaches while Fixed-Size Chunking (XC) has the lowest deduplication ratio. Note that XC works well on the LNX, WEB, VMB datasets, because LNX and WEB datasets have many files smaller than the fixed-size chunk of 10 KB (and thus the average generated chunk

Table 3.5 Comparison of deduplication ratio achieved by the five chunking approaches

Dataset	RC w/Min2KB	FC w/Min2KB	FC w/Min4KB	FC w/Min8KB	XC 10 KB
TAR	47.58%	47.64%	50.19%	47.18%	12.21%
LNX	97.25%	97.26%	97.35%	97.10%	96.51%
WEB	95.09%	94.02%	95.47%	94.44%	93.19%
VMA	38.23%	37.96%	40.31%	38.15%	18.26%
VMB	96.13%	96.09%	96.24%	96.11%	95.68%
RDB	95.53%	95.50%	96.71%	95.70%	9.80%
SYN	93.64%	93.67%	94.09%	92.62%	75.06%

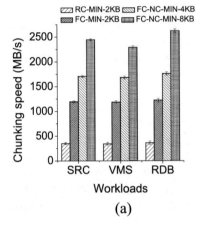
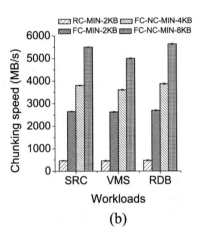

(a)   (b)

Fig. 3.15 Chunking speed of the four CDC approaches. (**a**) Speed on intel i7-930. (**b**) Speed on intel i7-4770

size also smaller than 10 KB) and VMB has many structured backup data (and thus VMB is suitable for XC).

Table 3.5 shows that RC and FC with Min2KB and XC generate similar average chunk size, while FC-NC with Min4KB has a slightly small average chunk size. But the approach of FC-NC with Min8KB has a much smaller average chunk size, which means that it generates fewer chunks and thus less metadata for deduplication processing. Meanwhile, FC-NC with Min8KB still achieves a comparable deduplication ratio, slightly lower than RC, while providing a much higher chunking speed as discussed later.

Figure 3.15 suggests that FC-NC with Min8KB has the highest chunking speed, about 10× faster than the Rabin-based approach, about 2× faster than FC with Min2KB. This is because FC-NC with Min8KB is the final FastCDC using all the three techniques to speed up the CDC process. In addition, FC-NC with Min4KB is also a good CDC candidate since it has the highest deduplication ratio while also working well on the other two metrics of chunking speed and the average generated chunk size. Note that XC is not shown here because it has almost no computation overhead for chunking.

Table 3.6 Number of instructions, instructions per cycle (IPC), and CPU cycles required to chunk 1 MB data by the four CDC approaches on the Intel i7-4770 processor

Approaches	Instructions	IPC	CPU cycles
RC-MIN-2KB	38, 829, 037	2.35	16, 537, 973
FC-MIN-2KB	15, 074, 950	4.37	3, 452, 146
FC-NC-MIN-4KB	11, 008, 372	4.82	2, 284, 453
FC-NC-MIN-8KB	7, 750, 124	4.82	1, 608, 033

Table 3.6 further studies the CPU overhead among the four CDC approaches. The CPU overhead is averaged on 1000 test runs by the Linux tool "Perf". The results suggest that FC-NC-MIN-8 KB has the fewest instructions for CDC computation, the highest IPC (instructions per cycle), and thus the least CPU time overhead, i.e., CPU cycles. Generally, FastCDC greatly reduces the number of instructions for CDC computation by using the techniques of Gear-based hashing and optimizing hash judgment (i.e., "FC-MIN-2KB"), and then minimizes the number of computation instructions by enlarging the minimum chunk size for cut-point skipping and combining normalized chunking (i.e., "FC-NC-MIN-8KB"). In addition, FastCDC increases the IPC for the CDC computation by well pipelining the instructions of hashing and hash-judging tasks in up-to-date processors. Therefore, these results explain why FastCDC is about $10\times$ faster than Rabin-based CDC, which is that the former not only reduces the number of instructions but also increases the IPC for the CDC process.

In summary, as shown in Tables 3.4–3.6, and Fig. 3.15, FastCDC (i.e., FC-NC-MIN-8KB) significantly speeds up the chunking process and achieves a comparable deduplication ratio while reducing the number of generated chunks by using a combination of the three key techniques.

3.4   Concluding Remarks

In this chapter we list the five key properties that a desirable CDC algorithm should have and discuss the problems facing the existing CDC algorithms. To address the problems, we present AE, a fast CDC algorithm with high throughput and small chunk-size variance thus high deduplication efficiency, for traffic redundancy elimination, and FastCDC, an extremely fast CDC algorithm for deduplication storage system. Results from our prototype evaluation driven by real-world datasets show the effectiveness and efficiency of AE and FastCDC.

Chapter 4
Indexing Schemes

Abstract Indexing is a key process to identify and eliminate duplicate data for deduplication systems. One of the main challenges is the scalability of fingerprint-index-based search schemes. The fingerprint space is too large to be stored in the memory. Access throughput of the on-disk fingerprint-index is too low to be acceptable for storage services. There are two primary approaches to scaling data deduplication: locality-based acceleration of deduplication and similarity-based deduplication. The rest of this chapter is organized as follows: Section 4.1 presents the correlated indexing schemes. Section 4.2 describes the problems and observations. Section 4.3 proposes the design and workflow of SiLo. Section 4.4 presents the performance evaluation on real-world datasets with the existing schemes.

4.1 Correlated Techniques of Indexing Scheme

Despite recent progress in data deduplication studies[16, 68, 71], many challenges remain, particularly in the petabyte-scale deduplication-based storage systems that are generally centralized. One of the main challenges is the scalability of fingerprint-index-based search schemes [16]. For example, to store a unique dataset of 1 PB and assuming an average chunk size of 8 KB, at least 2.5 TB of SHA-1 fingerprints will be generated, which are too large to be stored in the memory. State-of-the-art deduplication systems [16, 17, 68] suggest that the access throughput of the on-disk fingerprint-index is about 1–6 MB/s, which is too low to be acceptable for backup or storage services. Thus fingerprinting indexing has become the main performance bottleneck of large-scale data deduplication systems.

In order to address this performance bottleneck, many approaches have been proposed to improve the performance of deduplication indexing, by putting the hot fingerprints into RAM to minimize accesses to on-disk index and improve the throughput of deduplication. There are two primary approaches to scaling data deduplication: locality-based and similarity-based deduplication, such as Data Domain File System (DDFS) [16] and Extreme Binning respectively [71].

© Springer Nature Singapore Pte Ltd. 2022
D. Feng, *Data Deduplication for High Performance Storage System*,
https://doi.org/10.1007/978-981-19-0112-6_4

4.1.1 DDFS: Data Domain File System

At the highest level, DDFS divides a file into variable-length chunks and computes a fingerprint for each chunk. DDFS uses fingerprints to identify duplicate chunks and as a tag used to reference a chunk. In Fig. 4.1, DDFS [16] leverages a summary vector to reduce the number of times for accessing disk, which performs duplicate checking. If the summary vector indicates that a chunk is not in the index, the chunk is new and should be stored. On the other hand, if the summary vector indicates that the chunk is in the index, there is a high probability that the chunk is actually in the segment index, but there is no guarantee. Containers are self-describing in that a metadata section includes the segment descriptors for the stored segments.

DDFS employs a locality cache and a chunk prefetching solution to maintain the locality of the fingerprints. Moreover, it can remove 99% of the disk accesses for deduplication, which significantly improves the index throughput. However, it does not work in real-world datasets with poor locality. And DDFS consumes large amounts of memory and has a poor system scalability.

4.1.2 Extreme Binning

Extreme Binning splits up the chunk index into two tiers: primary index and bins. The primary index contains a representative fingerprint entry per file. The rest of the file's chunk fingerprints are kept on disk in the second tier that we call bin. Each representative fingerprint in the primary index contains a pointer to its bin. This two-tier index has been depicted in Fig. 4.2 [71]. Extreme Binning chooses the min fingerprint as the representative fingerprint per file, which is demonstrated by Broder's theorem [72].

Fig. 4.1 The system architecture and key structures of Data Domain File System (DDFS)

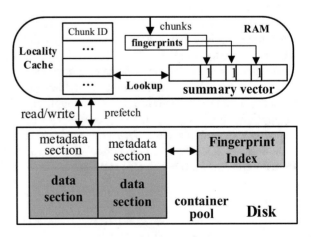

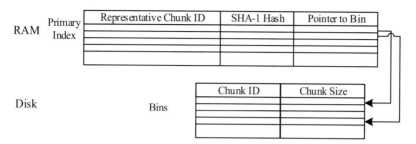

Fig. 4.2 The two-tier index structure and description of the primary index and bins of extreme binning

For a file, it will be divided into chunks, and its representative fingerprint and whole file hash is computed. The primary index is queried to find out if the file's representative fingerprint already exists in it. If not, a new secondary index or bin is created. All unique chunk IDs of the file along with their chunk sizes are added to this bin. The representative fingerprint, the whole file hash and a pointer to this newly created bin is added to the primary index. If the file's representative fingerprint is found in the primary index, its whole file hash is compared with the whole file hash in the primary index for the representative fingerprint. If the whole file hashes do not match, the bin pointer in the primary index is used to load the corresponding bin for duplicate checking.

Extreme Binning exploits data similarity to improve the index throughput and reduce the memory usage. However, it is a near-exact deduplication index scheme and will damage deduplication ratio for poor similarity datasets.

4.2 Performance Bottleneck and Exploiting File Characteristics

4.2.1 Low Throughput and High RAM Usage

The existing index approaches may result in low throughput and incur high RAM us-age. Specifically, locality-based approaches exploit the inherent locality in a backup stream, which is widely used in state-of-the-art deduplication systems such as DDFS [16], Sparse Indexing [68], and ChunkStash [73]. The locality in this context means that the chunks of a backup stream will appear in approximately the same order in each full backup with a high probability. Mining this locality increases the RAM utilization and reduces the accesses to on-disk index, thus alleviating the disk bottleneck.

Similarity-based approaches are designed to address the problem encountered by locality-based approaches in backup streams that either lack or have very weak locality (e.g., incremental backups). They exploit data similarity instead of locality in a backup stream, and reduce the RAM usage by extracting similar characteristics from the backup stream. A well-known similarity-based approach is Extreme Binning [71], which improves deduplication scalability by exploiting the file similarity to achieve a single on-disk index access for chunk lookup per file.

While these scaling approaches have significantly alleviated the disk bottleneck in data deduplication, there are still substantial limitations that prevent them from reaching the peta- or exa-scale, as explained below. Based on our analysis of experimental results, we find that in general a locality-based deduplication approach performs very poorly when the backup stream lacks locality while a similarity-based approach underperforms for a backup stream with a weak similarity. Unfortunately, the backup data in practice are quite complicated in how or whether locality/ similarity is exhibited.

4.2.2 Characteristics of the File Size

Our experimental observations, as well as intuition, suggest that the deduplication of large files can be very important while the deduplication of small files can be very time- and RAM-space consuming.

Large files. A typical file system contains many large files (e.g., ≥ 2 MB) that only account for less than 20% of total number of files but occupy more than 80% of the total space [4, 74], such as VMware images and database files. A recent study also suggests that the files larger than 1 GB account more than 90% of the total space in backup storage systems [6], because of backup software that tends to group individual files into "tar-like" collections. Obviously, these large files are an important consideration for a deduplication system due to their high space-capacity and bandwidth/time requirements in the inline backup process. The larger the files, the less similar they will be, even if significant parts within the files may be similar or identical, because the similarity-based approaches may miss the identification of significant redundant data in large files.

To address this problem of large files, SiLo approach divides a large file into many small segments to better expose similarity among large files[117]. More specifically, the probability that large files S_1 and S_2 share the same representative fingerprint is highly dependent on their similarity degree according to Broder's theorem [72, 75]: *Consider two sets S_1 and S_2, with $H(S_1)$ and $H(S_2)$ being the corresponding sets of the hashes of the elements of S_1 and S_2 respectively, where H is chosen uniformly and randomly from a min-wise independent family of permutations. Let min(S) denote the smallest element of the set of integers S. Then:*

$$Pr[\,min\,(H(S_1)) = \,min\,(H(S_2))] = \frac{|S_1 \cap S_2|}{|S_1 \cup S_2|} \tag{1}$$

This probability can be increased by segmenting the large files and detecting the similarity of all the segments of the large files, as follows. As files S_1 and S_2 are segmented into segments $S_{11} \sim S_{1n}$ and $S_{21} \sim S_{2n}$ respectively, similarity detection between S_1 and S_2 is determined by the union of the probabilities of similarity detection between $S_{11} \sim S_{1n}$ and $S_{21} \sim S_{2n}$. Based on the above probability analysis, this segmenting approach will only fail in the worst-case scenario where all the segments in file S_1 are not similar to segments of file S_2. This, based on the inherent locality in the backup streams, happens with a very small probability because it is extremely unlikely that two files are very similar but none of their respective segments is detected as being similar.

$$\begin{aligned} Pr[\,min\,(H(S_1)) = \,min\,(H(S_2))] &= \frac{|S_1 \cap S_2|}{|S_1 \cup S_2|} \\ &\ll Pr[\,min\,(H(S_{1l})) = \,min\,(H(S_{2l})) \cup \cdots \cup \,min\,(H(S_{1n})) = \,min\,(H(S_{2n}))] \\ &= \bigcup_{i=1}^{n} Pr[\,min\,(H(S_{1i})) = \,min\,(H(S_{2i}))] \\ &= 1 - \bigcap_{i=1}^{n} Pr[\,min\,(H(S_{1i})) \neq \,min\,(H(S_{2i}))] = 1 - \prod_{i=1}^{n}\left(1 - \frac{|S_{1i} \cap S_{2i}|}{|S_{1i} \cup S_{2i}|}\right) \end{aligned} \tag{2}$$

Small files. A file system typically contains a very large number of small files [4, 74]. Since the small files (e.g., $\leq$64 KB) usually only take up less than 20% of the total space of a file system but account for more than 80% of the total number of files, the chunk-lookup index for small files will be disproportionably large and likely out of memory. Consequently, the inline deduplication [16, 68] of small files will tend to be very slow and inefficient. This problem of small files can be addressed by grouping many highly correlated small files into a segment. We consider the logically adjacent files within the same parent directory to be highly correlated and thus similar. We exploit similarity and locality of a group (i.e., segment) of adjacent small files rather than one individual file or chunk. As a result, at most one access to on-disk index is needed per segment instead of per file or per chunk.

4.3   Design and Implementation of SiLo

In this section, we will describe the design, data structure, and workflow of SiLo as follows. The deduplication storage system consists of client and deduplication server (DS). The client is an interface for reading, chunking, and computing chunk fingerprints. We focus on Deduplication Server, which leverages the index for duplicate

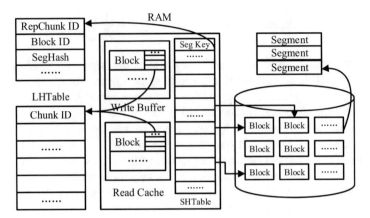

Fig. 4.3 Data structures of the deduplication server

checking. Deduplication server is the most likely performance bottleneck of the entire deduplication system. We present SiLo, which is a near-exact deduplication index for backup systems that exploits data locality and similarity to achieve high throughput and consume low memory overheads.

Deduplication server consists of Locality Hash Table (LHTable), Similarity Hash Table (SHTable), Write Buffer, and Read Cache. While SHTable and LHTable index segments and blocks, the similarity and locality units of SiLo respectively, the Write Buffer and Read Cache preserve the similarity and locality of backup streams, as shown in Fig. 4.3. In addition, RepChunkID is the representative fingerprint of a segment and Chunk ID is the SHA-1 fingerprint of a chunk. As mentioned above, the notion of segment is used to exploit the similarity of backup stream, while the block preserves the stream-informed locality layout of segments on the disk. SHTable provides the similarity detection for input segments and LHTable serves to quickly index and filter out duplicate chunks. The Write Buffer and Read Cache contain the recently accessed blocks to exploit the backup stream locality [16].

For the input data stream, SiLo will first use its similarity algorithm to pack correlated small files and divide large files into segments and check with SHTable to detect similarity. Then SiLo will use its locality algorithm to enhance the similarity detection thus find more duplicate data. Finally, SiLo will distribute the data-block to different storage nodes by a locality-based stateless routing algorithm.

Note that, since this chapter mainly aims to improve the indexing performance by making highly efficient use of the cache space of RAM and reducing accesses to on-disk fingerprints in the deduplication system, all write/read operations of segments/blocks in this chapter are performed in the form of writing/reading chunks' fingerprints while operations on the data-block are performed on the real backup data.

4.3.1 Similarity Algorithm

SiLo improves deduplication index scalability by combined exploitation of similarity and locality. It exploits similarity by grouping strongly correlated small files and segmenting large files, while locality is exploited by grouping contiguous segments in a backup stream to preserve the locality layout of these segments as depicted in Fig. 4.4. Thus, segments are the atomic building units of a block that is in turn the atomic unit of Write Buffer and Read Cache, and blocks are indexed by the unique block ID over the lifetime of the system.

As a key contribution of SiLo, the SiLo similarity algorithm is implemented in File Daemon, which structures data from backup streams into segments after the processing of chunking and fingerprinting according to the following three principles.

- P1. Fingerprint set of correlated small files in a backup stream (e.g., those under the same parent directory) are to be grouped into a segment.
- P2. Fingerprint set of a large file in a backup stream is divided into several independent segments.
- P3. All segments are of approximately the same size (e.g., 2 MB).

Where, P1 aims to reduce the RAM overhead of index-lookup; P2 helps expose more similarity characteristics of large files to eliminate more duplicate data; and P3 simplifies the management of segments. Thus, the similarity algorithm exposes and then exploits more similarity by leveraging file semantics and preserving locality-layout of a backup stream to significantly reduce the RAM usage for deduplication indexing.

The method of representative fingerprinting [71] is employed in SiLo to represent each segment by a similarity-index entry in the Similarity Hash Table. By virtue of P1, the SiLo similarity design solves the problem of small files taking up

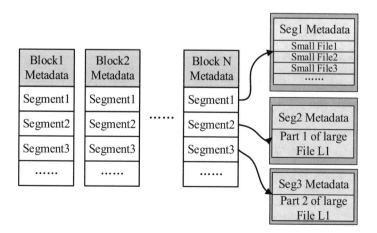

Fig. 4.4 Data structures of the SiLo similarity algorithm

disproportionably large RAM space. For example, assuming an average segment size of 2 MB and an average chunk or small file size of 8 KB, a segment accommodates 250 chunks or small files, thus significantly reducing the required index size in the memory. If we assume a 60-byte primary key for the similarity indexing of a 4 MB segment (backup data), which is considered economic, a 1 PB backup stream only needs 15 GB similarity-index for deduplication that can easily fit in the memory. Therefore, SiLo is able to use a very small and proper portion of RAM to support PB-scale deduplication by virtue of its similarity algorithm.

4.3.2   Locality Approach

As another salient feature of SiLo, the SiLo locality algorithm groups several contiguous segments in a backup stream into a block and preserves their locality-layout on the disk. The methods and workflow of this locality algorithm are depicted in Fig. 4.5. It helps detect more potentially duplicate chunks that are missed by the similarity detection. "N" refers to the fact that the segment is detected as dissimilar. According to the locality characteristic of backup streams, if input segment S_{1i} in block B_1 is determined to be similar to segment S_{2k} by hitting in Similarity Hash Table, SiLo will consider the whole block B_1 to be similar to block B_2 that contains S_{2k}. As a result, this grouping of contiguous segments into a block can eliminate more potentially duplicate data that is missed by the probabilistic similarity detection, thus complementing the similarity detection.

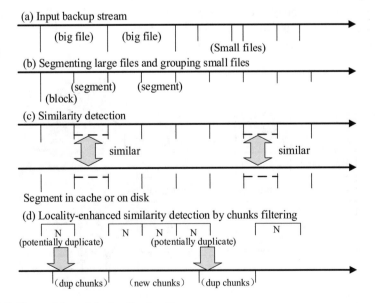

Fig. 4.5 The workflow of the locality algorithm

Since the block is the minimal write/read unit of Write Buffer and Read Cache in the SiLo system, it serves to maximize the RAM utilization and reduce frequent accesses to on-disk index by retaining access locality of the backup stream. When SiLo reads the blocks from disk by the similarity detection, it puts the recently accessed block into the Read Cache. By preserving the backup-stream locality in the Read Cache, the accesses to on-disk index due to similarity detection can be significantly reduced, which alleviates the disk bottleneck and increases the deduplication throughput.

The block size is an important system parameter that affects the system performance in terms of duplicate elimination and throughput. The smaller the block size, the more disk accesses will be required by the server to read the index, weakening the locality exploitation. The larger the block size, on the other hand, the more unrelated segments will be read by the server from the disk, increasing system's space and time overheads for deduplication indexing. Therefore, a proper block size not only provides good duplicate elimination but also achieves high throughput and low RAM usage in the SiLo system.

Each block in SiLo has its own Locality Hash Table (i.e., LHTable shown in Fig. 4.3) for chunk filtering. Since a block contains several segments, it needs an indexing tool for thousands of fingerprints. The fingerprints in a block are organized into the LHTable when reading the block from the disk. The additional time required for constructing LHTable in a block is significantly compensated by its quick indexing.

Beside the locality of Read Cache, we also exploit the locality of Write Buffer for data deduplication. Since users of file systems tend to duplicate files or directories under the same directories, a significant amount of duplicate data can be also eliminated by detecting the duplication in Write Buffer that also preserves the locality of a backup stream. For example, a code directory may include many versions of source code files or documents that become good deduplication candidates.

In our current design of SiLo, the Read Cache, and Write Buffer each contains a fixed number of blocks. Only a very small portion of RAM is thus used as the Write Buffer and Read Cache to store a small number of recently accessed blocks to avoid the frequent and expensive disk read/write operations. As illustrated in Figs. 4.3 and 4.4, a locality-block contains only metadata information such as LHTable, segment information, chunk information, and file information, which enables a 1 MB locality-block to represent a 200 MB data-block.

4.3.3   SiLo Workflow

To put things together and in perspective, Fig. 4.6 shows the main workflow of SiLo deduplication processes. Files in the backup stream are first chunked, fingerprinted, and packed into segments by grouping strongly correlated small files and

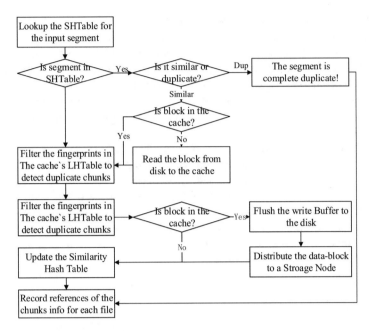

Fig. 4.6 The workflow of SiLo deduplication

segmenting large files in the File Agent. For an input segment S_{new}, SiLo goes through the following key steps:

- Check to see if S_{new} is in the SHTable. If it hits in SHTable, SiLo checks if the block B_{bk} containing S_{new}'s similar segment is in the cache. If it is not in the cache, SiLo will load B_{bk} from the disk to the Read Cache according to the referenced block ID of S_{new}'s similar segment, where a block is replaced in the FIFO order if the cache is full.
- The duplicate chunks in S_{new} are detected and eliminated by checking the fingerprint sets of S_{new} with LHTable (fingerprints index) of B_{bk} in the cache.
- If S_{new} misses in SHTable, it is then checked against recently accessed blocks in the read cache for potentially similar segment (i.e., locality-enhanced similarity detection).

Then SiLo will construct input segments into blocks to retain access locality of the input backup stream. For an input block B_{new}, SiLo does the following:

- The representative fingerprint of B_{new} will be examined to determine the stored backup nodes of data-block B_{new}.
- SiLo checks if the Write Buffer is full. If the Write Buffer is full, a block there is replaced in the FIFO order by B_{new} and then written to the disk.

After the process of deduplication indexing, SiLo will record the chunk-to-file mapping information as the reference for each file, which is managed by the Job Metadata of the Backup Server. For the read operation, SiLo will read the referenced metadata of each target file in the Job Metadata that allows the corresponding data chunks to be read from the data blocks in the Storage Server. These data chunks will then be used to reconstruct the target files in the File Daemon according to the index mapping relationship between files and deduplicated data chunks.

4.4 Performance Evaluation

We conduct our performance evaluation of SiLo on a platform of standard server configuration to evaluate and compare the inline deduplication performances of SiLo, ChunkStash, and Extreme Binning approaches running on a Linux environment. The hardware configuration includes a quad-core CPU running at 2.4 GHz, with a 4 GB RAM, 2 gigabit network interface cards, and two 500 GB 7200rpm hard disks.

Due to our lack of access to the source code of either the ChunkStash or Extreme Binning scheme, we have chosen to implement both of them. More specifically, we have implemented the locality-based and exact-deduplication approach of ChunkStash incorporating the principles and algorithms described in the ChunkStash paper [73]. The ChunkStash approach makes full use of the inherent locality of backup streams and uses a novel data structure called Cuckoo hash for fingerprint indexing. We have also implemented a simple version of the Extreme Binning approach, which represents a similarity-based and approximate-deduplication approach according to the algorithms described in the Extreme Binning paper [71]. Extreme Binning exploits file similarity instead of locality in the backup streams.

Note that our evaluation platform is not a production-quality deduplication system but rather a research prototype. Hence, our evaluation results should be interpreted as an approximate and comparative assessment of the three systems above, and not be used for absolute comparisons with other deduplication systems. The RAM usage in our evaluation is obtained by recording the memory allocated for index-lookup. The duplicate elimination performance metric is defined as the percentage of duplicate data eliminated by the system. Throughput of the system is measured by the rate at which fingerprints of the backup stream are processed, not the real backup throughput in that it does not measure the rate at which the backup data is transferred and stored.

Five traces representing different strengths of locality and similarity are used in the performance evaluation of the three deduplication systems and are listed in Table 4.1. The five traces are collected from real-world datasets of One-backup, Incremental-backup, Linux-version, and two Full-backup sets respectively. All use SHA-1 for chunk fingerprints and the Content-Defined Chunking algorithm. The deduplication factor is defined as the Totalsize/(Totalsize-Dedupsize) ratio.

Table 4.1 Features of the five traces used in the performance evaluation

Feature	One-set	Inc-set	Linux	Full-set1	Full-set2
Total size	530 GB	251 GB	101 GB	2.51 TB	6.0 TB
Total files	3.5 M	0.59 M	8.8 M	11.3 M	12.1 M
Total chunks	51.7 M	29.4 M	16.9 M	417.6 M	1.87 G
Average chunk size	10 KB	8 KB	5.9 KB	6.5 KB	4.9 KB
Dedup factor	1.7	2.7	19	25	16.7
Locality	Weak	Weak	Strong	Strong	Strong
Similarity	Weak	Strong	Strong	Strong	Strong

- One-set trace was collected from 15 graduate students of our research group. Since we obtain only one full backup for this group, this trace has weak locality and weak similarity.
- Inc-set is a subset of the trace reported by Tan et al. [76] and was collected from initial full backups and subsequent incremental backups of eight members in a research group. There are 391 backups with a total of 251 GB data. Since the operation of incremental backup only backs up the modified and new files after the first full backup, Inc-set represents datasets with strong similarity but weak locality.
- Linux-set, downloaded from the website [77], consists of 900 versions from version 1.1.13 to 2.6.33, and represents the characteristics of small files.
- Full-set1 consists of 380 full backups of 19 researchers' PCs over 20 days, which is reported by Xing et al. [78]. Full-set1 represents the datasets with strong locality and strong similarity.
- Full-set2 was collected from an engineering group consisting of 15 graduate students and was used in [79]. The students in this group ran full or incremental backups independently in a span of 31 days. Full-set2 also represents datasets with strong locality and strong similarity.

Linux-set, Full-set1, and Full-set2 are used in the previous studies [71, 78, 79] respectively to evaluate the performance of Extreme Binning, and our use of these datasets resulted in similar and consistent evaluation results with the published studies.

4.4.1 Duplicate Elimination

Figure 4.7 shows the duplicate elimination performance of the three systems under the five workloads. Since ChunkStash does the exact deduplication, it eliminates 100% of duplicate data. Compared with Extreme Binning that eliminates 71–99% of duplicate data in the five datasets, SiLo removes about 98.5–99.9% of duplicate data. Note that, while Extreme Binning eliminates about 99% of duplicate data as expected in Linux-set, Full-set1, and Full-set2 that has strong similarity and locality, it fails to detect almost 30% of duplicate data in One-set that has weak locality and

Fig. 4.7 Comparison among ChunkStash, SiLo, and Extreme Binning in terms of percentage of duplicate data eliminated on the five datasets

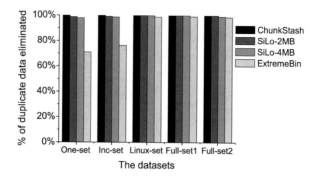

Fig. 4.8 Comparisons among ChunkStash, SiLo, and Extreme Binning in terms of RAM usage (B: RAM required per MB backup data)

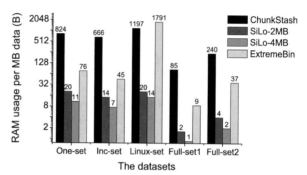

similarity, and about 25% of duplicate data in Inc-set with weak locality but strong similarity. Although there is strong similarity in Inc-set, Extreme Binning still fails to eliminate a significant amount of duplicate data primarily due to its probabilistic similarity detection that simply chooses one representative fingerprint for each file regardless of the file size.

On the contrary, SiLo-2 MB eliminates 99% of duplicate data even in One-set with both weak similarity and locality, and also removes almost 99.9% of duplicate data in Linux-set, Full-set1, and Full-set2 with both strong similarity and locality. These results show that SiLo's joint and complementary exploitation of similarity and locality is very effective in detecting and eliminating duplicate data under all workloads, achieving near-complete duplicate elimination.

4.4.2   RAM Usage for Deduplication Indexing

Figure 4.8 shows the RAM usage for deduplication among these three systems under the five workloads. For Linux-set that has a very large number of small files and small chunks, the highest RAM usage is incurred for both Chunkstash and Extreme Binning. There is also a clear negative correlation between the deduplication factor and the RAM usage for the approximate deduplication systems of SiLo and Extreme Binning on the other four workloads. That is, for One-set that has the lowest

Table 4.2 RAM usage in a PB-scale deduplication system for different deduplication approaches

State-of-the-art dedupe approaches	Exact or approximate	Average chunk size	RAM usage per PB data
DDFS	Exact	8 KB	125 GB
Sparse Indexing	Approximate	4 KB	85 GB
Extreme Binning	Approximate	N/A	300 GB
ChunkStash	Exact	8 KB	0.75 TB
MAD2	Exact	4 KB	1 TB
HPDS	Approximate	4 KB	50 GB
SiLo-2 MB	Approximate	N/A	30 GB
SiLo-4 MB	Approximate	N/A	15 GB

deduplication factor, the highest RAM usage is incurred, while for Full-set1 that has the highest deduplication factor, the smallest RAM space is required.

The average RAM usage for ChunkStash is the highest among the three approaches, except for the Linux-set trace, as it does the exact deduplication that needs a large hash table in the memory to put all the indices of chunk fingerprints. Although ChunkStash uses the Cuckoo hash to store the compact key signatures instead of full chunk-fingerprints, it still requires at least 6 bytes for each new chunk. In addition, according to the open-source code of Cuckoo Hash which is used in this chapter for ChunkStash evaluation [80], it needs to allocate about 2 million slots in advance to support 1 million index entries. Note that using the variant of Cuckoo hash may incur a high load factor but still store at least 6 bytes for each new chunk in the ChunkStash system.

Since only the file similarity index needs to be stored in the RAM, Extreme Binning only consumes about 1/9–1/15 of the RAM space required of ChunkStash, except on the Linux-set where it consumes more RAM usage than ChunkStash due to the extremely large number of small files. However, SiLo-2 MB's RAM efficiency allows it to reduce the RAM consumption of Extreme Binning by a factor of 3–900. The extremely low RAM overhead of SiLo stems from the interplay between its similarity and locality algorithm. On the other hand, the RAM usage for Extreme Binning depends on the average file size of datasets, in addition to the deduplication factor. The smaller the average file size is, the more RAM space Extreme Binning will consume, which is demonstrated in the Linux-set. The RAM usage of SiLo remains relatively stable with the change in average file size in the five traces and is inversely proportional to the deduplication factor of the traces as shown in Fig. 4.8.

Table 4.2 shows the RAM usage in a PB-scale deduplication system for several state-of-the-art approaches. We assume that the average file size is 200 KB, secure fingerprint is SHA-1, and chunks are not compressed. As a representative fingerprint requires about 60 bytes as key index in the memory regardless of the average chunk size, Extreme Binning [71] demands almost 300 GB of RAM space with a mean file size of about 200 KB [74] in a PB-scale deduplication system while SiLo-2 MB and SiLo-4 MB consume about 30 GB and 15 GB memory by a similarity- and locality-based deduplication index design. DDFS [16] consumes about 125 GB RAM space for deduplicating 1 PB unique data (one byte per chunk by the Bloom filter) while Sparse Indexing [68] reduces the RAM usage to about 85 GB by a sparse index de-sign with 1/64 sampling. HPDS [23] further reduces the RAM usage to about 50 GB by a progressive sampled index with a sampling rate of 1/101. Both ChunkStash [73] and MAD2 [79] consume almost 1 TB of RAM space to maintain a global index in a PB-scale deduplication system by their Cuckoo-hash-based and Bloom-filter-array-based indexing schemes respectively. Thus SiLo uses significantly less RAM space than the above state-of-the-art approaches. As the RAM space is still limited in computer systems, the lower usage of memory means a higher scalability of the SiLo deduplication system.

4.4.3 Deduplication Throughput

Figure 4.9 shows a comparison between the three approaches in terms of deduplication throughput, where the throughput is observed to be more than double, as the average chunk size changes from 5 KB (e.g., Full-set2) to 10 KB (e.g., One-set). ChunkStash achieves an average throughput of about 292 MB/s with a range of 24–654 MB/s on the five datasets. The frequency of accesses to on-disk index by ChunkStash's compact key signatures algorithm on the Cuckoo hash lookup tends to increase with the size of the dataset, thus adversely affecting the throughput. Extreme Binning achieves an average throughput of 768 MB/s with a range of 158–1571 MB/s on the five datasets, since it only needs to access the disk once per similar-file and eliminates the duplicate files in the memory. As SiLo-2 MB makes at most one disk access per segment, it deduplicates data at an average throughput of 1042 MB/s with a range of 538–1486 MB/s on the five datasets.

Although Extreme Binning runs faster than SiLo-2 MB under Inc-set where many duplicate files exist, it runs much slower in other datasets. Since each bin tends to grow in size with the size of datasets, Extreme Binning will slow down as the size of each bin increases. This is because each similar file must read its corresponding bin

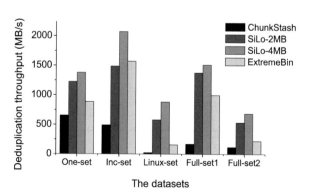

Fig. 4.9 Comparison between ChunkStash, SiLo, and Extreme Binning in terms of deduplication throughput (MB/s)

in its entirety. In addition, the design of bin fails to exploit the backup-stream locality that helps reduce disk accesses and increase the RAM utilization by preserving the locality layout in RAM.

Therefore, compared with Extreme Binning and ChunkStash, SiLo is shown to provide robust and consistently good deduplication performance, achieving higher throughput, and near-exact duplicate elimination at a much lower RAM overhead.

4.5   Concluding Remarks

This chapter shows the disk access bottleneck of the index for large-scale deduplication systems. It describes that the fingerprint index is too large to store in the memory, and storing fingerprints in the disk will lead to low-throughput challenges. The current locality-based and similarity-based index schemes, for example, DDFS and Extreme Binning, fail to solve the problems on the real-world datasets. This chapter presents SiLo, which exploits both similarity and locality in backup streams to achieve higher deduplication throughput, and near-complete duplicate elimination with an extremely lower RAM overhead for large-scale systems. Results from our evaluation show that the SiLo achieves high throughput and low memory overheads.

Chapter 5
Rewriting Algorithms

Abstract Data restore is an important phase in the storage management stage of data deduplication systems. During a restore, a recipe (i.e., the fingerprint sequence of a backup) is read, and the containers serve as the prefetching unit. After deduplication, the chunks of each backup are often physically scattered, also known as fragmentation problem, which severely decreases restore performance. In this chapter, we discuss the problem and present the solution to alleviate the problem. The rest of this chapter is organized as follows: Section 5.1 presents the state-of-the-art algorithms for addressing the fragmentation problem. Section 5.2 describes the design and implementations of our proposed History-Aware Rewriting algorithm for data center. Section 5.3 describes our another causality-based deduplication performance booster for both cloud backup and restore operations.

5.1 Development of Defragmentation Algorithm

In deduplication-based backup systems, the chunks of each backup are physically scattered after deduplication, which causes a challenging fragmentation problem. The fragmentation decreases restore performance, and results in invalid chunks becoming physically scattered in different containers after users delete backups. Figure 5.1 illustrates an example of two consecutive backups to show how the fragmentation arises. The shaded areas in each container represent the chunks required by the second backup. There are 13 chunks in the first backup. Each chunk is identified by a character, and duplicate chunks share an identical character. Two duplicate chunks, say A and D, are identified by deduplicating the stream, which is called self-reference. A and D are called self-referred chunks.

All unique chunks are stored in the first 4 containers, and a blank is appended to the 4th half-full container to make it be aligned. We observe that the second backup contains 13 chunks, 9 of which are duplicates in the first backup. The 4 new chunks are stored in 2 new containers. However, all chunks in the first backup are stored in the first 3 containers in order, while all chunks are scattered across 6 containers for the second backup. This phenomenon is referred to as fragmentation. Meanwhile, as shown in Fig. 5.1, with a 3-container-sized LRU cache, restoring the first backup

© Springer Nature Singapore Pte Ltd. 2022
D. Feng, *Data Deduplication for High Performance Storage System*,
https://doi.org/10.1007/978-981-19-0112-6_5

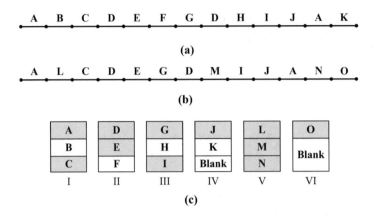

Fig. 5.1 An example of two consecutive backups. (**a**) First backup stream. (**b**) Second backup stream. (**c**) Data layout after second backup stream

needs to read 5 containers. The self-referred chunk A requires extra reading container I. With a 3-container-sized LRU cache, restoring the second backup needs to read 9 containers. Although both of the backups consist of 13 chunks, restoring the second backup needs to read 4 more containers than restoring the first backup. Hence, deduplication causes fragmentation, which exacerbates restore performance.

To solve the fragmentation problem, Nam et al. propose a quantitative metric to measure the fragmentation level of deduplication systems, and a selective deduplication scheme for backup workloads. The Context-Based Rewriting algorithm (CBR) and the capping algorithm (CAP) are recently proposed to rewrite duplicate but fragmented chunks to address the fragmentation problem. iDedup eliminates sequential and duplicate chunks in the context of primary storage systems. SAR maintains the chunks with a high reference count in SSD to accelerate the restore speed. However, how to design a defragmentation algorithm for the deduplication-based backup system still faces challenges in terms of improving both the restore performance and the deduplication ratio.

Chunk Fragmentation Level. Chunk fragmentation level (CFL) has been proposed as an effective indicator of restore performance [64, 87]. CFL proposes a quantitative metric to measure the fragmentation level of deduplication systems, and a selective deduplication scheme for backup workloads. Specifically, while a calculated fragmentation level is too high, only deduplicate sequences of duplicate chunks (ignoring any intermixed non-duplicate chunks) are stored in the same chunk container if the sequence is long enough. In the presence of caching, this is suboptimal because it penalizes switching back to a recently used container.

Context-Based Rewriting Algorithm. Context-based rewriting algorithm (CBR) uses a fixed-sized buffer, called stream context, to maintain the following chunks of the pending duplicate chunk that is being determined whether fragmented. Specifically, CBR defines the rewrite utility of a pending chunk as the size of the chunks that are in the disk context (physically adjacent chunks) but not in the stream context, divided by the size of the disk context. If the rewrite utility of the pending chunk is higher than the predefined minimal rewrite utility, the chunk is fragmented. There are two fixed-size contexts of a duplicate—its disk context and stream context. The stream context of a block in a stream is defined as a set of blocks written in this stream immediately after this block, whereas its disk context contains blocks immediately following this block on disk. When the intersection of these two contexts is substantial, reading of blocks in this intersection is very fast due to prefetching. Thus, CBR avoids too many rewrites in a limited buffer.

Capping Rewriting Algorithm. Capping rewriting algorithm (CAP) divides the backup stream into 20MB fixed-sized segments, and conjectures the fragmentation within each segment. In a limited buffer, CAP ranks order the containers by how many chunks of the segment they contain, and chooses the top T containers which contain the most chunks. Suppose a new segment refers to N containers and N>T, the chunks in the N-T containers that hold the least chunks in the segment are rewritten.

iDedup: Inline Deduplication. iDedup deduplicates a sequence of duplicate blocks only when its corresponding already stored blocks are sequentially laid out on disk and exceed a minimum length threshold. Specifically, we derived two insights by observing real-world, primary workloads: (i) there is significant spatial locality on disk for duplicated data, and (ii) temporal locality exists in the accesses of duplicated blocks. First, iDedup leverages spatial locality to perform deduplication only when the duplicate blocks form long sequences on disk, thereby, avoiding fragmentation. Second, iDedup leverages temporal locality by maintaining dedup-metadata in an in-memory cache to avoid extra IOs. Although iDedup eliminates sequential and duplicate chunks in the context of primary storage systems, it limits how often seeks must be done when sequentially reading back data.

SAR: SSD-Assisted Restore. SAR is an efficient deduplication-based storage system with an improved restore performance. Specifically, SAR stores the unique data chunks with high reference count, small size, and nonsequential characteristics in SSDs, which exploits the high random-read performance properties of SSDs and the unique data sharing characteristic. SAR monitors and identifies the chunk size, sequentiality and the reference counts of the unique data chunks via a "Selective Promotion" module. And SAR also records access popularity of the unique data chunks. SAR selectively stores the "hot" unique data chunks in SSD array based on the above information.

Based on the approaches above, the existing defragmentation algorithms still introduce many problems, which motivates us to propose new defragmentation algorithms to eliminate fragmentation and satisfy the need for different scenarios. Thus, we proposed the History-Aware Rewriting algorithm for identifying fragmentation accurately [139]. Meanwhile, we propose a Causality-Based Deduplication (CABdedup) performance booster for cloud backup services [145].

5.2   HAR: History-Aware Rewriting Algorithm

5.2.1   Fragmentation Classification

The fragmentation comes from two containers, the sparse container and the out-of-order containers. In this subsection, we describe two types of containers, and then present our key observations that motivate this work.

Sparse Container. As shown in Fig. 5.1, only one chunk in container IV is referenced by the second backup. Prefetching container IV for chunk J is inefficient when restoring the second backup. After deleting the first backup, we require a merging operation to reclaim the invalid chunks in container IV. This kind of containers exacerbates system performance on both restore and garbage collection. We define a container's utilization for a backup as the fraction of its chunks referenced by the backup. If the utilization of a container is smaller than a predefined utilization threshold, such as 50%, the container is considered as a sparse container for the backup. We use the average utilization of all the containers related with a backup to measure the overall sparse level of the backup.

Sparse containers directly amplify read operations. Prefetching a container of 50% utilization at most achieves 50% of the maximum storage bandwidth, because 50% of the chunks in the container are never accessed. Hence, the average utilization determines the maximum restore performance with an unlimited restore cache. The chunks that have never been accessed in sparse containers require the slots in the restore cache, thus decreasing the available cache size. Therefore, reducing sparse containers can improve the restore performance.

After backup deletions, invalid chunks in a sparse container fail to be reclaimed until all other chunks in the container become invalid. Symantec reports the probability that all chunks in a container become invalid is low. We also observe that garbage collection reclaims little space without additional mechanisms, such as offline merging sparse containers. Since the merging operation suffers from a performance problem similar to the restore operation, we require a more efficient solution to migrate valid chunks in sparse containers.

Out-of-Order Container. If a container is accessed many times intermittently during a restore, we consider it as an out-of-order container for the restore. As shown in Fig. 5.1, container V will be accessed three times intermittently while restoring the second backup. With a 3-container-sized LRU restore cache, restoring each chunk in container V incurs a cache miss that decreases restore performance.

The problem caused by out-of-order containers is complicated by self-references. The self-referred chunk D improves the restore performance, since the two accesses to D occur close in time. However, the self-referred chunk A decreases the restore performance.

The impacts of out-of-order containers on restore performance are related to the restore cache. For example, with a 4-container-sized LRU cache, restoring the three chunks in container V incurs only one cache miss. For each restore, there is a minimum cache size, called cache threshold, which is required to achieve the maximum restore performance (defined by the average utilization). Out-of-order containers reduce restore performance if the cache size is smaller than the cache threshold. They have no negative impact on garbage collection.

A sufficiently large cache can address the problem caused by out-of-order containers. However, since the memory is expensive, a restore cache of larger than the cache threshold can be unaffordable in practice. Hence, it is necessary to either decrease the cache threshold or assure the demanded restore performance if the cache is relatively small. If restoring a chunk in a container incurs an extra cache miss, it indicates that other chunks in the container are far from the chunk in the backup stream. Moving the chunk to a new container offers an opportunity to improve restore performance. Another more cost-effective solution to out-of-order containers is to develop a more intelligent caching scheme than LRU.

5.2.2 Inheritance of Sparse Containers

Because out-of-order containers can be alleviated by the restore cache, how to reduce sparse containers becomes the key problem. Existing rewriting algorithms cannot accurately identify sparse containers due to the limited buffer. Accurately identifying sparse containers requires the complete knowledge of the on-going backup. However, the complete knowledge of a backup cannot be known until the backup has concluded, making the identification of sparse containers a challenge.

Due to the incremental nature of backup, two consecutive backups are very similar, which is the major assumption behind DDFS. Hence, they share similar characteristics, including the fragmentation. We analyze three datasets, namely virtual machines, Linux kernels, and a synthetic dataset, to explore and exploit potential characteristics of sparse containers (the utilization threshold is 50%). After each backup, we record the accumulative amount of the stored data, as well as the total and emerging sparse containers for the backup. An emerging sparse container is not sparse in the last backup but becomes sparse in the current backup. An inherited sparse container is already sparse in the last backup and remains sparse in the current backup. The total sparse containers are the sum of emerging and inherited sparse containers.

The characteristics of sparse containers are shown in Fig. 5.2, and 100 backups are shown for clarity. First, the number of total sparse containers continuously grows. It indicates sparse containers become more common over time. Second, the number of total sparse containers increases smoothly most of the time. A few exceptions in the Kernel datasets are major revision updates, which have more new data and increase the amount of stored data sharply. It indicates that a large update results in more emerging sparse containers. However, due to the similarity between consecutive backups, the number of emerging sparse containers of each backup is relatively small most of the time. Third, the number of inherited sparse containers of each backup is equivalent to or slightly less than the number of total sparse containers of the previous backup. A few sparse containers of the previous

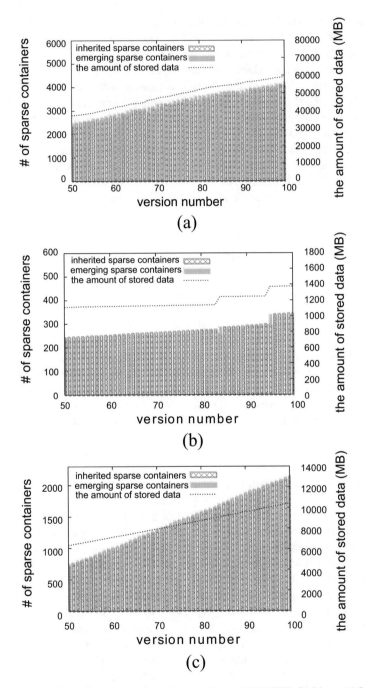

Fig. 5.2 Characteristics of sparse containers in three datasets. (**a**) VMDK. (**b**) Linux. (**c**) Synthetic

backup become not sparse to the current backup since their utilizations drop to 0. It seldom occurs that the utilization of an inherited sparse container increases in the current backup, unless a rare rollback occurs. The observation indicates that sparse containers of the backup remain sparse in the next backup.

The above observations motivate our work to exploit the historical information to identify sparse containers. After completing a backup, we can determine which containers are sparse within the backup. Because these sparse containers remain sparse for the next backup, we record these sparse containers and allow chunks in them to be rewritten in the next backup. In such a scheme, the emerging sparse containers of a backup become the inherited sparse containers of the next backup. Due to the second observation, each backup needs to rewrite the chunks in a small number of inherited sparse containers, which would not degrade the backup performance. Moreover, a small number of emerging sparse containers left to the next backup would not degrade the restore performance of the current backup. From the third observation, the scheme identifies sparse containers accurately. This scheme is called History-Aware Rewriting algorithm (HAR) [139].

5.2.3   History-Aware Rewriting Algorithm

Figure 5.3 illustrates the overall architecture of our HAR system. On disks, we have a container pool to provide container storage service. Any kind of fingerprint index can be used. Typically, we keep the complete fingerprint index on disks, as well as the hot part in memory. An in-memory container buffer is allocated for chunks to be written. The system assigns each dataset a globally unique ID, such as DS1 in Fig. 5.3. The collected historical information of each dataset is stored on disks with the dataset's ID, such as the DS1 info file. The collected historical information consists of three parts: IDs of inherited sparse containers for HAR, the container-

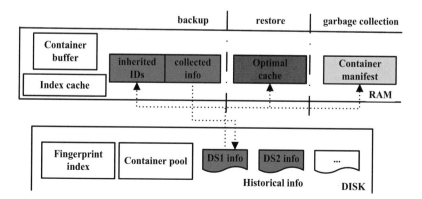

Fig. 5.3 The HAR architecture

access sequence for the Belady's optimal replacement cache, and the container manifest for Container-Marker Algorithm.

At the beginning of a backup, HAR loads IDs of all inherited sparse containers to construct the in-memory $S_{\text{inherited}}$ structure, and rewrites all duplicate chunks in the inherited sparse containers. In practice, HAR maintains two in-memory structures, S_{sparse} and S_{dense} (included in collected info in Fig. 5.3), to collect IDs of emerging sparse containers. The S_{sparse} traces the containers whose utilizations are smaller than the utilization threshold. The S_{dense} records the containers whose utilizations exceed the utilization threshold. The two structures consist of utilization records, and each record contains a container ID and the current utilization of the container. After the backup is completed, HAR replaces the IDs of the old inherited sparse containers with the IDs of emerging sparse containers in S_{sparse}. Hence, the S_{sparse} becomes the $S_{\text{inherited}}$ of the next backup. The complete workflow of HAR is described in Algorithm 1.

Algorithm 1: History-Aware Rewriting Algorithm

Require: Input: IDs of inherited sparse containers, $S_{inherited}$;
Require: Output: IDs of emerging sparse containers, S_{sparse};
  1: Initialize two sets, S_{sparse} and S_{dense}.
  2: **while** the backup is not completed **do**
  3:     Receive a chunk and look up its fingerprint in the fingerprint index.
  4:     **if** the chunk is not duplicate **then**
  5:         **if** the chunk's container ID exists in $S_{inherited}$ **then**
  6:             Rewrite the chunk, and obtain a new container ID.
  7:         **else**
  8:             Eliminate the chunk.
  9:         **end if**
 10:     **else**
 11:         Write the chunk, and obtain a new container ID.
 12:     **end if**
 13:     **if** the chunk's container ID doesn't exist in S_{dense} **then**
 14:         Update the associated utilization record (add it if doesn't exist) in S_{sparse} with the chunk size.
 15:         **if** the utilization exceeds the utilization threshold **then**
 16:             Move the utilization record to S_{dense}.
 17:         **end if**
 18:     **end if**
 19: **end while**
 20: **return** S_{sparse}

5.2.4 *Optimal Restore Cache*

To reduce the negative impacts of out-of-order containers on restore performance, we implement Belady's optimal replacement cache. Implementing the optimal cache (OPT) needs to know the future access pattern. We can collect such information during the backup, since the sequence of reading chunks during the restore is just the same as the sequence of writing them during a backup.

After a chunk is processed through either elimination or overwriting its container ID, its container ID is known. We add an access record into the collected info in Fig. 5.3. Each access record can only hold a container ID. Sequential accesses to the identical container can be merged into a record. This part of historical information can be updated to disks periodically, and thus would not consume much memory.

At the beginning of a restore, we load the container-access sequence into memory. If the cache is full, we evict the cached container that will not be accessed for the longest time in the future. Belady has proven the optimality.

The complete sequence of access records can consume considerable memory when out-of-order containers are dominant. Assuming each container is accessed 50 times intermittently and the average utilization is 50%, the complete sequence of access records of a 1TB stream consumes over 100MB of memory. Instead of checking the complete sequence of access records, we can use a slide window to check a fixed-sized part of the future sequence, as a near-optimal scheme. The memory footprint of this near-optimal scheme is hence bounded. Because the recent backups are most likely restored, we only maintain the sequences of a few recent backups for storage savings, and restore earlier backups via an LRU replacement caching scheme.

5.2.5 *A Hybrid Scheme*

Rewriting chunks in out-of-order containers offers opportunities to reduce their negative impacts. Since most of the chunks rewritten by existing rewriting algorithms belong to out-of-order containers, we propose a hybrid scheme that takes advantages of both HAR and existing rewriting algorithms (e.g., CBR and CAP) as optional optimizations. The hybrid scheme is straightforward. Each duplicate chunk not rewritten by HAR is further examined by CBR or CAP. If CBR or CAP considers the chunk fragmented, the chunk is rewritten.

To avoid a significant decrease of deduplication ratio, we configure CBR or CAP to rewrite less data than the exclusive uses of themselves. For example, CBR uses a rewrite limit to control the rewrite ratio (the size of the rewritten chunks divided by that of the total chunks). The default rewrite limit in CBR is 5\%, and thus CBR attempts to rewrite top-5\% fragmented chunks. Generally, a higher rewrite limit indicates CBR rewrites more data for higher restore performance. We set rewrite limit to 0.5\% in the hybrid of HAR and CBR. The hybrid of HAR and CAP is similar. Based on our observations, only rewriting a small number of additional chunks further improves restore performance when the restore cache is small. However, the hybrid scheme always rewrites more data than HAR. Hence, we propose disabling the hybrid scheme if a large restore cache is affordable (Since restore is rare and critical, a large cache is reasonable).

5.2.6   Performance Evaluation

We implemented an experimental platform to evaluate our design, including HAR, OPT, and CMA. To better illustrate the problem, we use the state-of-the-art rewriting algorithms CBR, CAP, and their hybrid schemes (HAR+CBR and HAR+CAP) for comparisons.

Experimental Configurations. We implemented an experimental platform to evaluate our design, including HAR, OPT, and CMA. We also implement CBR (The original CBR is designed for HydraStor, and we implement the idea in the container storage), CAP, and their hybrid schemes (HAR+CBR and HAR+CAP) for comparisons. Since the design of fingerprint index is out of scope for the chapter, we simply accommodate the complete fingerprint index in memory. The baseline has no rewriting, and the default caching scheme is OPT. The container size is 4MB. The default utilization threshold in HAR is 50\%. We retain 20 backups, thus backup n-20 is deleted after backup n is finished. We do not apply the offline container merging as in previous work, because it requires a long idle time.

We use speed factor as the metric of the restore performance. The speed factor is defined as 1 divided by mean containers read per MB of restored data. Higher speed factor indicates better restore performance. Given the container size is 4 MB, 4 units of speed factor correspond to the maximum storage bandwidth.

Datasets. Two real-world datasets, namely VMDK and Linux, and a synthetic dataset, i.e., Synthetic, are used for evaluation. Their characteristics are listed in Table 5.1. Each dataset is divided into variable-sized chunks.

VMDK is from a virtual machine installed Ubuntu 12.04LTS, which is a common use-case in real-world. We compile source code, patch the system, and run an HTTP server on the virtual machine. We backup the virtual machine regularly. It consists of 102 full backups. Each full backup is 14.48 GB on average, and 90\$–\$98\%

Table 5.1 Characteristics of datasets

Dataset name	VMDK	Linux	Synthetic
Total size	1.44 T	104 GB	4.5 TB
# of versions	102	258	400
Deduplication ratio	25.44	45.24	37.26
Avg. chunk size	10.33 KB	5.29 KB	12.44 KB
Sparse	Medium	Severe	Severe
Out-of-order	Severe	Medium	Medium

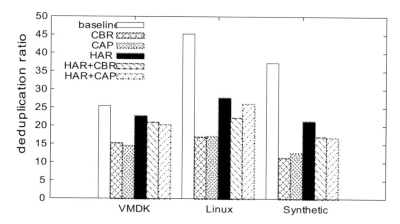

Fig. 5.4 The comparisons between HAR and other rewriting algorithms in terms of deduplication ratio

identical to its adjacent backups. Each backup contains about 15% self-referred chunks, and thus out-of-order containers are dominant.

Linux, downloaded from the web, is a commonly used public dataset. It consists of 258 consecutive versions of unpacked Linux kernel sources. Each version is 412.78 MB on average. Two consecutive versions are generally 99% identical except when there are large upgrades. In Linux, there are only a few self-references and sparse containers are dominant.

Synthetic is generated according to existing approaches. We simulate common operations of file systems, such as create/delete/modify files. We finally obtain a 4.5 TB dataset with 400 versions. There is no self-reference in Synthetic.

Deduplication Ratio. Deduplication ratio explains the amount of written chunks, and the storage cost if no backup is deleted. Figure 5.4 shows deduplication ratios of rewriting algorithms. The deduplication ratios of HAR are 22.78, 27.78, and 21.38 in VMDK, Linux, and Synthetic respectively. HAR rewrites 11.66%, 62.83%, and 74.31% more data than the baseline. However, the corresponding rewrite ratios remain at a low level, respectively 0.45%, 1.38%, and 1.99%. It indicates the size of rewritten data is small relative to the size of backups. Due to such low rewrite ratios, the fingerprint lookup, content-defined chunking, and SHA-1 computation remain the performance bottleneck. Meanwhile, HAR achieves considerably higher deduplication ratios than CBR and CAP. Since the rewrite ratios of CBR and CAP are two times larger than that of HAR, it is reasonable to expect that HAR outperforms CBR and CAP in terms of backup performance. The hybrid schemes, HAR +CBR and HAR+CAP, achieve better deduplication ratio than CBR and CAP respectively, but decreased deduplication ratios compared to HAR, such as by 10% in VMDK.

Restore Performance. Figure 5.5 shows the restore performance achieved by each rewriting algorithm with a given cache size. The cache is 512-, 32-, and 64-container-sized in VMDK, Linux, and Synthetic respectively. We tune the cache size

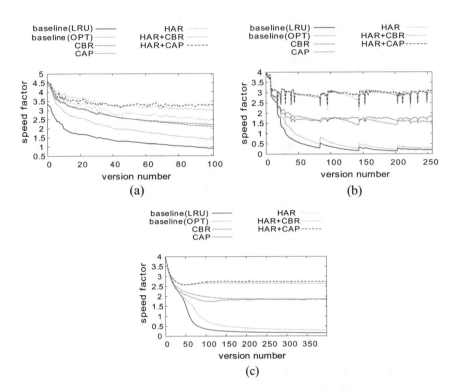

Fig. 5.5 The comparisons of rewriting algorithms in terms of restore performance. (**a**) VMDK. (**b**) Linux. (**c**) Synthetic

according to the datasets, and show the impacts of varying cache size later in Fig. 5.6. Speed factor is the average value of last 20 backups. The cache size is in terms of sharp of containers. The default caching scheme is OPT. We observe severe declines of the restore performance in the baseline. For instance, restoring the latest backup is 21X slower than restoring the first backup in Linux. OPT alone increases restore performance by 1.51X, 1.47X, and 1.88X respectively in last 20 backups; however, the performance remains at a low level.

We further examine the average speed factor in last 20 backups of each rewriting algorithm. In VMDK, CBR and CAP further improve restore performance by 1.46X and 1.53X respectively based on OPT. HAR outperforms them and increases restore performance by a factor of 1.72. The hybrid schemes are efficient, because HAR +CBR and HAR+CAP increase restore performance by 1.2X and 1.3X based on HAR. Given that their deduplication ratios are slightly smaller than HAR, CBR and CAP are good complements to HAR in the datasets where out-of-order containers are dominant. The restore performance of the initial backups exceeds the maximum storage bandwidth (4 units of speed factor), because self-referred chunks in the scope of the cache improve restore performance.

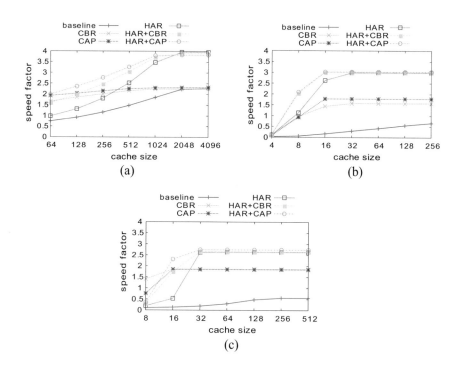

Fig. 5.6 The comparisons of rewriting algorithms under various cache size. (**a**) VMDK. (**b**) Linux. (**c**) Synthetic

In Linux, CBR and CAP further improve restore performance by 5.4X and 6.12X. HAR is more efficient and further increases restore performance by a factor of 10.25. Because out-of-order containers are less dominant, the hybrid schemes cannot achieve significantly better performance than HAR. Thus the hybrid schemes can be disabled in the datasets where the problem of out-of-order containers is less severe. There are some occasional smaller values in the curve of HAR, because a large upgrade in Linux kernel produces a large amount of sparse containers. The results in Synthetic are similar to those in Linux. CBR, CAP, and HAR further increase restore performance by 6.41X, 6.35X, and 9.08X respectively. The hybrid schemes cannot outperform HAR remarkably.

Figure 5.6 compares restore performance among rewriting algorithms under various cache sizes. In VMDK, because out-of-order containers are dominant, HAR requires a large cache (e.g., 2048-container-size) to achieve the maximum restore performance. We observe that if the cache size continuously increases, the restore performance of the baseline is approximate to that of CBR and CAP. The reason is that the baseline, CBR, and CAP achieve similar average utilizations as shown in Fig. 5.6. CBR and CAP are great complements to HAR. When the cache is small, the restore performance of HAR+CBR (HAR+CAP) is approximate to that of CBR(CAP); when the cache is large, the restore performance of the hybrid schemes is approximate to that of HAR. Compared with HAR, the hybrid schemes success-fully decrease the cache threshold by nearly 2X, and improve the restore perfor-mance when the cache is small.

In Linux, HAR achieves better restore performance than CBR and CAP, even with a small cache (e.g., 8-container-size). Compared to HAR, the hybrid schemes decrease the cache threshold by a factor of 2, and improve the restore performance when the cache is small. However, because the cache threshold of HAR is small, a restore cache of reasonable size can address the problem caused by out-of-order containers without decreasing deduplication ratio. In Synthetic, HAR outperforms CBR and CAP by 1.41X and 1.42X when the cache is no less than 32-container-size. With a small cache (e.g., 8-container-size), CBR and CAP are better. However, because the cache threshold of HAR is small, it is reasonable to allocate sufficient memory for a restore. The hybrid schemes improve restore performance when the cache is small.

5.3   A Causality-Based Deduplication Performance Booster

5.3.1   File Causality of Backup Datasets

In both the traditional backup and cloud backup environments, there are two critical performance metrics, backup window (BW) and recovery time objective (RTO), to evaluate the backup and recovery performances respectively. Backup window represents the time spent on sending specific datasets to the backup destination while recovery time objective denotes the maximum amount of downtime a business is willing to accept after data disasters. A recent ESG (i.e., Enterprise Strategy Group) research has indicated that about 58\% of professionals in SMBs can tolerate no more than 4 h of downtime before experiencing significant adverse effect. This will be a much bigger challenge for cloud backup services due to the relatively low bandwidth of WAN (Wide-Area Network) that underpins the cloud backup platform. Particularly, in cloud backup environments, many products, such as EMC Avamar, Cumulus, Asigra, and Commvault Simpana, have adopted the source-side data deduplication technology to reduce the backup time by removing redundant data from transmission to backup destinations. However, despite the critical importance of the restore time in achieving a reasonable RTO as discussed earlier, much less attention has been paid to reducing the time spent on restoring data from remote backup destinations to the user's local computer for cloud recovery. In this subsec-tion, we discuss the causality-induced data redundancy in existing backup systems that provides a potential opportunity for us to remove redundant data from data transmission and present our key observations that motivate this work.

Causality-Induced Data Redundancy. In backup systems, there are many backed-up versions of the same dataset stored at the backup destination due to multiple full and incremental backups. Except for the initial full backups, the dataset backed up each time is evolved from its previous backed-up versions with data modifications.

Fig. 5.7 Successive backups and restores illustrated in directed graphs. (**a**) Successive three backups at times t1, t2, and t3. (**b**) Restore to the backup point at time t3 from time t4

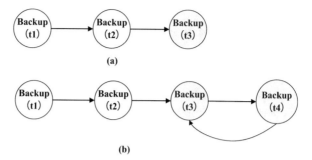

As a result, it is the modified data, rather than the unmodified data that is already stored at the backup destination by previous backed-up versions, that is required to be transmitted each time. The same is true during the data restores. Each restore operation takes place after data corruptions and needs to restore the corrupted dataset to a previous backed-up version stored at the backup destination. Therefore, during either data backup or restore operations, it is possible to reduce the amount of data transmitted over WAN by removing from transmission the unmodified data shared between the current dataset being processed (i.e., dataset to be backed up or restored each time) and its previous backed-up versions. In what follows, we present the causal connections among files in multiple versions of the same dataset to reveal the prevalent existence of unmodified data during each backup/restore operation.

- **Unchanged Files.** Either in the directory backups or directory restores, there are many files kept intact after their last backups.
- **Modified Files.** In most cases, the individual files that require data backups or restores have always been changed with data modifications, insertions, or deletions after their backups.
- **Deleted Files.** Besides the unchanged and modified files, typically there are some files that have been deleted after their last backups.

The three cases described above intuitively describe the causal relationships among the different versions of the same dataset in backup/restore scenarios, suggesting that a large amount of unmodified data can be identified and removed from transmissions to reduce the total amount of data transmitted.

Mining Data Redundancy. We introduce an alternative data redundancy mining model to analyze the removal of the unmodified data that exists among multiple backup and restore operations as mentioned before. To clearly present the data redundancy mining model, we use a Directed Graph (DG) to show several successive data backups and restores of the same dataset in Fig. 5.7. Figure 5.7(a) shows three successive backup operations that run at t_1, t_2, and t_3 ($t_1 < t_2 < t_3$) respectively, in which the backup at t_1 is the initial full backup of this dataset. Figure 5.7(b) shows a successive restore operation at time t_4 after three preceding backups, which tries to restore the corrupted dataset to the backup point t_3. This data redundancy mining model quantifies the amount of the data that is required to be transmitted for each backup/restore operation after removing its unmodified data from transmission.

Above studies observe that that a large amount of redundant data exists among multiple data backups and restores. Most datasets processed in both data backups and data restores are evolved from their previous backed-up versions with relatively minor modifications, insertions, or deletions, resulting in most files and data chunks unchanged in their entirety after backups. The observations and analyses motivate us to propose a Causality-Based deduplication performance booster for cloud backup services, called CABdedupe [145]. By capturing and preserving this causal relationship among chronological versions of datasets, it is possible to fast identify which files have been changed and which data chunks differ among multiple file versions, thus helping remove the unmodified data from transmission to significantly reduce the total transferred dataset in both backup and restore operations.

5.3.2   CABdedup Architecture

In this section, we present the system architecture of CABdedupe and describe how it can be applied to the existing cloud backup systems. As shown in Fig. 5.8, the assumed general backup system consists of two software components: Client and Server. Client, installed on the user's local computer, is responsible for sending/ retrieving the backup dataset to/from the backup destination, while Server, located in service provider's data center, is responsible for storing/returning the backed-up dataset from/to Client. To succinctly illustrate the role of CABdedupe in a backup system, we use Backup-Client and Backup-Server to represent the functionalities of the original client and server modules in existing backup systems. CABdedupe consists of CAB-Client and CAB-Server.

CAB-Client. CAB-Client is composed of two functional modules: the Causality-Capture module and Redundancy-Removal module. The former, consisting of File Monitor, File List, and File Recipe Store, is responsible for capturing the causal

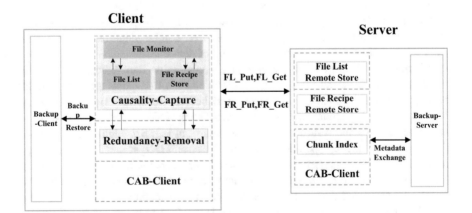

Fig. 5.8 System Architecture of CABdedupe

relationships among different files, while the latter is responsible for removing the unmodified data for each backup/restore operation with the help of the captured causality information by the former. Of the components of the Causality-Capture module, File Monitor is a daemon process that works at the file system level to keep track of some specific file operations, including file rename, file create, file delete, and file-content modification, File List is responsible for logging these file operations and File Recipe Store is responsible for saving the file recipes (i.e., the fingerprints of data chunks) of the backed-up files. These three components collectively capture the causal relationships among the different files in multiple backups and restores.

CAB-Server. CAB-Server, with File List Remote Store and File Recipe Remote Store as its components, stores the file lists and file recipes sent from CAB-Client, which ensures the availability of the causality information captured by CAB-Client in case of CAB-Client's corruptions. However, due to data transmission overheads, the file lists and file recipes stored in CAB-Server are not updated in the same timely fashion as that stored in CAB-Client.

Another component of CAB-Server, Chunk Index, is responsible for locating and retrieving the data chunks stored in CAB-Server for data restores. However, this Chunk Index component can be excluded from CAB-Server if the original backup system has exploited the chunk-level deduplication capability and its intended functionality already exists in Backup-Server.

Interface. In this subsection, we describe several interfaces that are used to integrate CABdedupe into the existing backup systems as showed in Fig. 5.8. These interfaces are classified into three categories as follows. The first interface, with Backup and Restore, is used to connect CAB-Client with Backup-Client. During each backup/restore operation, Backup-Client communicates with CAB-Client through this interface to remove the unmodified data from transmission by the Redundancy-Removal module. The second interface, with FL Put, FL Get, FR Put, and FR Get, is used to store the file lists and file recipes in CABServer. FL Put and FL Get are used to exchange the file lists, while FR Put and FR Get are used for the exchange of file recipes between CAB-Client and CAB-Server. The third interface, called Metadata Exchange, is responsible for the communication of chunk metadata information between CAB-Server and Backup-Server. The metadata information, including chunk fingerprints, chunk addresses, and chunk lengths, is used for building chunk index in CAB-Server during backups and locating the corresponding data chunks during restores. However, this interface will not be necessary if there is no Chunk Index in CAB-Server.

Redundancy Removal. In CABdedupe, the unmodified data is removed by the Redundancy-Removal module with the help of the causality information stored in File List and File Recipe Store, as follows.

1. *Backup:* During each backup operation, Backup-Client communicates with CAB-Client through the following four steps to remove its unmodified data.

   • Step 1. Check File List to find which files have been modified after their last backup.

- Step 2. For each modified file, check whether the file metadata and file content have been changed.
- Step 3. If some files have only file metadata modifications, CAB-Client notifies Backup-Client to only back up their modified file metadata instead of the whole file content. Otherwise, the files are chunked by the Rabin Fingerprints algorithm and each chunk is named by the SHA-1 hash function to filter out the unmodified data chunks in the next step.
- Step 4. This step finds and removes the unmodified data chunks by checking the file recipes of the previously backed-up versions. After filtering out the unmodified data chunks, CAB-Client notifies Backup-Client to back up the remaining modified data chunks to Backup-Server, and the file recipes of the new backed-up files versions are saved in File Recipe Store for redundancy exploitation in future backup/restore operations.

2. *Restore:* During each restore operation, CABdedupe takes the following four steps to remove its unmodified data, similar to the backup operation. The only difference between the backup and restore operations is that, during each backup operation, it removes the unmodified data chunks that have already been stored in Server by previous backups, while during each restore operation, the redundant data chunks removed are those kept intact in Client after their last backups (i.e., one backup point to be restored).

- Step 1. Get the file recipes of all the files in the restored dataset (i.e., one backed-up version).
- Step 2. For each file in the restored dataset, check File List to find its current file version existing in Client to see whether it has been changed or not after its last backup (i.e., the restored backup point).
- Step 3. If the files are kept intact after their backups, CAB-Client notifies Backup-Client that these file are not required to be retrieved from Backup-Server. Otherwise, the files are chunked by the Rabin Fingerprints algorithm and each chunk is named by the SHA-1 hash function to find the unmodified data chunks in the next step.
- Step 4. This step is to find and remove the unmodified data chunks by checking the file recipes of the files in the restored dataset. After filtering out these unmodified data chunks, CAB-Client notifies Backup-Client to retrieve the remaining modified data chunks from Backup-Server.

5.3.3 *Exploring and Exploiting Causality Information*

In CABdedupe, the causality information among different files plays a crucial role in removing the redundant data during the backup/restore operations. In this section, we present how CABdedupe uses its key components to capture and leverage the causality information.

Table 5.2 An example of the file list

File name	Revised name	Modification flag	Last backup time
. . .	. . .	. . .	. . .
/home/file/1.txt	–	MetaModify|ContentModify	01-13 15:07:34
/home/file/2.txt	–	MetaModify	01-13 15:07:34
/home/file/3.txt	/home/data/3.txt	Rename|MetaModify	01-13 15:07:34
/home/file/4.txt	/home/delete/4.tmp	Delete	01-13 15:07:34
–	/home/file/5.txt	Create|MetaModify|ContentModify	01-13 15:07:34
. . .	...	...	...

File Monitor. File Monitor is a daemon process running at the file system level to monitor the file-system-level system calls to keep track of some file operations. It is triggered when the user's system is bootstrapped and terminated when the system is shut down. To reduce its overhead, CABdedupe only captures a small portion of file operations so as to keep File Monitor idle or lightly loaded during most of the time.

File Set Consideration. CABdedupe only focuses on the directories and files that have been initially fully backed up, ignoring any other files existing in the same file system. Moreover, given that most files are small and holding a very small amount of data in typical file systems \cite{agrawal2007five}, CABdedupe excludes these small files to further reduce the size of the file set.

File Operation Consideration. CABdedupe captures only the first file write operation after each backup, to track whether the files (i.e., including file metadata and file content) have been modified, ignoring all the other file write and read operations. At the same time, two other special file operations, file rename (i.e., file content is not changed) and file delete, are also captured by CABdedupe to track the data redundancy. For the file deletions, CABdedupe renames the deleted files and makes them only visible to CABClient, thus preventing them from actual deletions in case of later file restores. However, restricted by the available storage space in Client, not all the files can be prevented from file deletions forever. When the space usage reaches a preset limit, CABdedupe uses a FIFO (First In First Out) replacement algorithm to reclaim the used space.

File List. File List (FL) is a table that is used to record the file operations captured by File Monitor. Table 5.2 depicts its four entries with an example.

- *Last Backup Time.* The time when the file was the most recently backed up.
- *Modification Flag.* This flag indicates the files' modification status after their last backups, including file create, file rename, file delete, file-metadata modification, and file-content modification. Note that the status of some successive modifications may be overlapped. For example, a file create operation creates both its metadata and content, thus resulting in both file-metadata modification and file-content modification.
- *Original File Name and New File Name.* These two entries are used to keep track of the file rename operations. The Original File Name represents the original name by which the file is backed up and the New File Name denotes the new file name after the rename operations.

The above four entries describe the causality information that is stored in File List. During each backup/restore operation, CABdedupe checks File List to find the files that have been modified since the most recent backups. In order to limit the size of File List so as to accelerate this search process, CABdedupe excludes the small files along with their file operations. In our experiments, we observed that 86.7% of files are smaller than 8KB during directory backups/restores, but these small files only occupy 2.6% of the total storage space. By excluding those files smaller than 8KB from the CABdedupe process, it can remove 86.7% of the files that the Causality-Capture module must otherwise consider while only failing to eliminate 0.54% of the redundant data. Moreover, CABdedupe can use the FIFO or LRU replacement algorithms, or delete the outdated entries in File List to further restrict the growth of its size.

Table 5.2 shows one example of File List that contains all of the file operations, including file create, file rename, file delete, file-metadata modification, and file-content modification, which CABdedupe mainly focuses on. In this table, the files "/home/file/1.txt" and "/home/file/2.txt" have been modified, file "/home/file/3.txt" has been renamed to file "/home/data/3.txt", file "/home/file/4.txt" has been deleted, and file "/home/file/5.txt" has been created after the backup of directory "/home" at 15:07:34 on January 13, 2010. During the next backup, the modified file metadata and file content of the files "/home/file/1.txt" and "/home/file/5.txt" had to be backed up, while for the files "/home/file/2.txt" and "/home/file/3.txt", only their modified file metadata needed to be backed up. During the next restore operation (i.e., restored to the backup point at 15:07:34 on January 13, 2010), for the files "/home/file/1.txt" and "/home/file/2.txt", retrieval of their file metadata and file content from the backup destination was required, while for files ``/home/file/3.txt" and "/home/file/4.txt", only their renaming was required. So according to the causality information stored in File List, CABdedupe can easily find out which files have been modified after their most recent backups.

File Recipe Store. File Recipe Store is a container used to store the file recipes. As showed in Fig. 5.9, each file recipe consists of two parts, the file metadata and data chunk list. The former contains file name, file backup time, and metadata chunk fingerprint, while the latter includes the chunk fingerprints and chunk sizes of all the data chunks that constitute a specific file. With the help of these file recipes, CABdedupe can easily locate the unmodified data chunks to be removed from transmission for each backup/restore operation.

Causality Information Consistency. During each backup/restore operation, CABdedupe relies heavily on the causality information that is stored in File List and File Recipe Store to identify and remove unmodified data chunks. Therefore, CABdedupe must ensure that the causality information stored in File List and File Recipe Store is consistent with that actually existing among the datasets from multiple backups/restores. The inconsistency of this information can lead to false positives in that unmodified data is regarded as modified data or false negatives in that modified data is regarded as unmodified data. Although the false positives cannot affect the routine backups/restores but can only degrade the effectiveness

Fig. 5.9 An example of file
recipe

```
File Metadata:
      File Name:/home/file/1.txt
      Backup Time:2010-01-13 15:07:34
      Metadata Chunks:8d7ks20t82…


Data Chunks List(Chunk Fingerprint, Chunk
Size(Bytes)):
      8616ef68Ac…,16186
      eb59eb2363...,6455
      612e7a35Ba...,7735
      5737588Aae...,12340
      03872e1Dcc...,7807

              ⋮
```

of CABdedupe, the false negatives can mislead the routine backups/restores to skip
the backups or restores of some modified data chunks.

To avoid this problem, CABdedupe only stores the consistent causality information to File List and File Recipe Store in CAB-Client. After each backup/restore
operation, CABdedupe buffers and accumulates the newly captured causality information until the next backup/restore operation, thus ensuring that the captured
information among these two backups/restores is accurate and consistent. Moreover,
CABdedupe stores this consistent causality information in both CAB-Server and
CAB-Client. When the causality information stored in CAB-Client is corrupt,
CABdedupe retrieves the same information from CAB-Server and rebuilds File
List and File Recipe Store to improve the next backup/restore performance. However, due to the lack of the new causal relationship between the dataset processed in
the next backup/restore and its previous backed-up versions in the newly built File
List and File Recipe Store, CABdedupe must scan the file system to find the
unmodified data chunks for the optimization of the next backup/restore operation.
This detailed process and the detailed maintenance of the causality information
consistency implemented by CABdedupe are both omitted in this chapter due to
space constraints.

5.3.4 Performance Evaluation

We have built our prototype systems and fed real-world data sets to evaluate the
performance of the CABdedupe.

Experimental Setup

1. *Prototype Systems*: We have integrated CABdedupe into two existing backup systems.

   - *Cumulus.* Cumulus is a cloud backup system \cite{mvrable1993cumulus} that exploits the source-side chunk-level deduplication approach to remove the redundant data from transmission for backup operations while ignoring the restore operations. We select it as a baseline system to assess how effective CABdedupe is in optimizing the backup/restore performances of an existing cloud backup system that already exploits the data deduplication technology. We denote the prototype system of Cumulus integrated with our CABdedupe as Cumulus+CAB. Cumulus+CAB has inherited all the intrinsic characteristics of Cumulus, such as supporting switches to different storage providers by storing CABdedupe's file lists and file recipes in normal files in providers' sites.

   - *MBacula.* As its name implies, MBacula is developed based on an open-source backup software called ``Bacula'' \cite{Bacula} with some modifications to Bacula (version 2.0.3), such as the data layout optimization, improving the flow control of backups/restores, and etc. However, MBacula does not exploit the deduplication technology to optimize both the backup and restore performances. We select it as another baseline system to assess how effective CABdedupe is in optimizing the backup/restore performances of an existing backup systems that does not exploit the data deduplication technology. We call the prototype system of MBacula integrated with our CABdedupe as MBacula+CAB. In MBacula+CAB, CABdedupe's components, such as File List, File Recipe Store, and Chunk Index, are implemented in a database form for fast queries and retrievals of the causality information.

   In these prototype systems, each client or server machine is featured with two-socket dual-core 2.1 GHz CPUs, a total of 2GB memory, 1 Gbps NIC cards, and a 500 GB hard drive.

2. *Datasets:* We report the experimental results based on three datasets with different characteristics. The first dataset, used to evaluate the directory backups/restores, is the full backups of one author's home directory lasting for one month, totaling about 275.17GB data. Its key statistics is showed in Table 5.3. The other two datasets, a database file generated in MBacula and a tar file of Linux source tree, both have five backed-up versions as described in Table 5.4, which are used to evaluate the restore performances of individual files.

Restore Performance. In this subsection, we show both the directory restore performance and individual file restore performance separately to analyze the impact of the restore optimizations by CABdedupe. However, because it is hard to trigger realistic data disasters, we inject data corruptions by developing a program that randomly selects some files to apply the file rename operations, file modifications (including data insertion, deletion, and modification) and file deletions to the last

Table 5.3 Key statistics of one author's home directory

Duration		31 days
The status of the dataset on the 31st day	Entries	31,647
	Files	28,837
	Average (file size)	314.5 KB
	Median (file size)	43.1 KB
	Maximum (file size)	253.6 MB
	Total (file size)	9.07 GB
Average update rates	New data/day	12.9 MB
	Changed data/day	39.7 MB
	Total data/day	52.6 MB

Table 5.4 The file sizes of the five backed-up versions of the database file and tar file

	File size (MB)	
File version	Database file	Tar file
1st version	6.06	294.06
2nd version	7.62	353.21
3rd version	8.38	365.83
4th version	9.32	382.42
5th version	10.07	394.91

Fig. 5.10 The amount of the redundant data removed and the corresponding transfer cost reduced by CABdedupe during each of the 30 restores

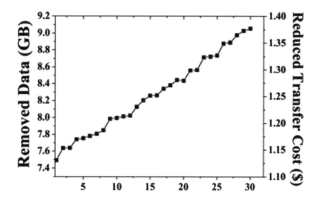

version of the dataset to simulate data disasters, so that the restore operations are invoked to restore the datasets to their previous backed-up versions to show the restore performances. While we realize that these are contrived data disasters, we have not been able to find a better way to simulate/emulate real-world disasters.

1. *Single Directory Restore Performance:* We use the datasets of 31 full backups of one author's home directory (i.e., not the whole file system) to evaluate the directory restore performance. There are a total of 30 simulated data restore operations by restoring this directory on the 31st day to its previous backed-up versions on the 1st day, 2nd day, 3rd day, ..., 30th day.

   Figure 5.10 shows the amount of the redundant data removed and the corresponding transfer cost reduced by CABdedupe during each of these

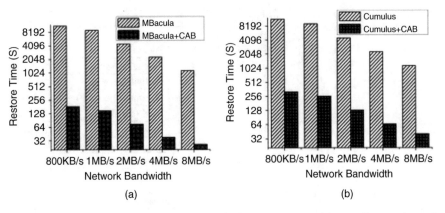

Fig. 5.11 Directory Restore times of MBacula and MBacula+CAB. (**a**) Directory Restore times of MBacula and (**b**) Directory Restore times of Cumulus and Cumulus+CAB MBacula+CAB

30 data restores. We adopt the transfer prices set for Amazon S3 in April 2010 as the pricing criteria, at about 0.15 dollars per GB. As shown in this figure, the amount of the removed data during each restore is gradually increasing as the restored backup point approaches the 31st day, implying that more data transfer cost can be reduced when restoring the directory to the more recent backup points.

To more accurately quantify the optimizations of the directory restore performance, we focus on the restore operation that restores the home directory on the 31st day to its backed-up version on the 30th day. During this restore operation, the redundant data that can be removed by CABdedupe is about 8.93GB, and the transfer costs that can be reduced by Cumulus+CAB and MBacual+CAB are about 1.34 and 1.32 respectively. From these results, we find that the reduced transfer cost by MBacual+CAB is proportional to the amount of the removed redundant data by CABdedupe. But this is not the case for Cumulus+CAB. This is because Cumulus+CAB groups many data chunks into one segment as a file stored in service provider's site as Cumulus does, and, as a result, retrieving a single data chunk requires the retrieval of the entire segment, followed by the extraction of the required data chunk from the segment. This results in many irrelevant data chunks being transmitted by Cumulus+CAB. We set the segment size to 16MB in both Cumulus+CAB and Cumulus.

Figure 5.11 compares the restore times of MBacula and MBacula+CAB, Cumulus and Cmulus+CAB, where we simulate a network environment with different network bandwidths: 800 KB/s, 1 MB/s, 2 MB/s, 4 MB/s, 8 MB/s. As the figure shows, both Cmulus+CAB and MBacula+CAB significantly reduce the restore times of their respective original (CAB-less) backup systems, in particular with a reduction ratio of 61.9 : 1 by MBacula+CAB under the network bandwidth of 800KB/s. This is because, during directory restores, most of the files are kept intact after their backups that require no data transmission, given that most data writes are centered on a small subset of files in typical file systems. On the other

Table 5.5 The amount of the redundant data removed from transmission by CABdedupe

File version	Database file	Tar file
1st version	6.05 MB	37.05 MB
2nd version	7.6 MB	63.28 MB
3rd version	8.37 MB	77.75 MB
4th version	9.3 MB	105.4 MB

Table 5.6 The amount of the data transmitted by MBacula+CAB and Cumulus+CAB during the individual file restores for restoring the 5th versions

Version	Database file (MB) (+CAB)		Tar file (MB) (+CAB)	
	MBacula	Cumulus	MBacula	Cumulus
1st	0.01	6.06	257.1	289.1
2nd	0.02	7.62	290.08	322
3rd	0.01	8.38	288.16	320.16
4th	0.02	9.32	277.11	321.12

hand, due to the transmission of many irrelevant data chunks by Cumulus and Cumulus+CAB, the restore times of both Cumulus and Cumulus+CAB are longer than that of MBacula and MBacula+CAB.

2. *Individual File Restore Performance:* Besides directory restores, the restore of individual files is another common restore operation in cloud backup environments. In most cases, the individual file restore happens when the file in the client site is changed or deleted, which is different from the directory restore where many files are kept intact after their backups. Thus during individual file restore operations, the amount of redundant data that can be removed is much less than that in directory restores. In our experiments, we select a database file generated in MBacula and a tar file of the Linux source tree as our datasets to show the individual file restore performances. Both the database and tar files have five backed-up versions as described in Table 5.4. We simulate four restore operations by restoring the 5th file version to the 1st, 2nd, ..., and 4th file version for each of the two files.

Table 5.5 shows the amount of the redundant data removed during each restore operation. It is during the individual file restore operations for restoring the 5th versions of database and tar files to their 1st, 2nd , ..., 4th versions. Similar to directory restores, the amount of the redundant data removed by CABdedupe increases as the file version approaches the 5th version. However, this common trend observed in both the directory restores and the individual file restores should not be regarded as a rule, since the amount of this redundant data is heavily dependent on the amount of data modifications to each specific dataset.

Table 5.6 compares the amount of the data required to be transmitted by MBacula +CAB and Cumulus+CAB during those restores presented in Table 5.5. From these results, we find that Cumulus+CAB transmits much more data than MBacula+CAB, especially in the restores of the database file. This is because in Cumulus+CAB, each backed-up version of the database file is smaller than 16 MB, and each time we have

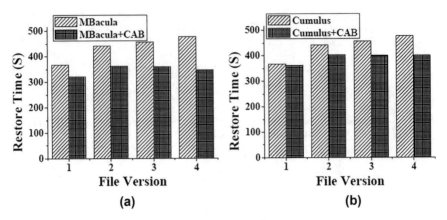

Fig. 5.12 The comparison of the individual file restore times. (**a**) restore times of MBacula and MBacula+CAB, and (**b**) restore times of Cumulus and Cmulus+CAB

grouped all the data chunks in one file stored in the service provider's site, thus resulting in no data reduction since any restore must retrieve the entire database file.

Figure 5.12 compares the restore times of MBacula and MBacula+CAB, Cumulus and Cumulus+CAB when restoring the 5th version of the tar file to its 1st, 2nd, ..., and 4th version respectively under the network bandwidth of 800 KB/s. This figure reveals that MBacula+CAB incurs 27% less restore time than MBacula, whereas the restore time of Cumulus+CAB is only 12% less than that of Cumulus due to the backup format stored in the service provider's site.

5.4   Concluding Remarks

This chapter describes the low restore performance caused by the fragmentation problems, and thus providing defragmentation methods is very necessary for deduplication-based storage systems. This chapter discusses the impact of the fragmentation caused by the deduplication, and then proposes the HAR and the CABdedup algorithm for backup system and cloud storage respectively. Specifically, HAR algorithm accurately identifies and rewrites sparse containers via exploiting historical information. Moreover, CABdedup algorithm also enables the removal of the unmodified data from transmission to improve restore performance with captured causality information.

Chapter 6
Secure Deduplication

Abstract Data privacy and confidentiality become concerns for encrypted deduplication storage systems. The attackers intend to obtain the secret information that do not belong to them by performing side-channel attack and unauthorized access. Users usually encrypt data with their own keys to ensure confidentiality of outsourced data, which prohibits cross-user deduplication. It becomes an urgent problem to develop a secure and efficient deduplication system on encryption and key management. The rest of this chapter is organized as follows: Section 6.1 presents the related work of secure deduplication schemes. Section 6.2 describes the problems and the motivation of SecDep. Section 6.3 proposes the system architecture, and the design of user-aware convergent and multi-level key management. Section 6.4 presents the security analysis under the proposed threat model. Section 6.5 presents the experiment evaluation.

6.1 Progress of Secure Deduplication

Harnik et al. [99] put forward that cross-user client-side deduplication can be exploited as a side-channel to obtain users' sensitive data, for example, bank account and password. Mulazzani et al. [100] and Dropship [101] discover hash manipulation attack on Dropbox to access unauthorized files. Furthermore, Halevi et al. [102] point out that an adversary who has the hash value can download data from CSP. To ensure outsourced data confidentiality, users usually encrypt data with their own keys, which prohibits cross-user deduplication. Specifically, the same data encrypted by different users' keys will result in different ciphertexts such that duplicates will not be found. However, sharing keys among users can result in data leakage because of user compromise [103]. Most of the existing secure deduplication solutions use a deterministic encryption method called convergent encryption (CE) [104] or message-locked encryption (MLE) [105, 106]. CE uses a hash value of the data as a key to encrypt the data. CE will encrypt the identical data into the same ciphertexts, which enables deduplication on the ciphertexts. However, building a secure and high-performance deduplication-based cloud backup system still faces challenges in terms of convergent encryption and key management.

© Springer Nature Singapore Pte Ltd. 2022
D. Feng, *Data Deduplication for High Performance Storage System*,
https://doi.org/10.1007/978-981-19-0112-6_6

6.1.1   Convergent Encryption or Message-Locked Encryption

Convergent encryption (CE) or message-locked encryption (MLE) is presented as four-tuple polynomial-time algorithms. HCE is similar with CE, expect for tag generation algorithm. The key of HCE is the hash value of the message and the tag is the hash value of the key. The algorithm definition of convergent encryption is shown as follows.

- GenKey(M) $\rightarrow$ K. K is a message-derived key that is generated by a key generation algorithm. A message (i.e., data) M is the input.
- Encry(P, K, M) $\rightarrow$ C. *Encry* is a symmetric encryption algorithm. The message-derived key K, and message M are inputs. C is the ciphertext.
- GenTag(P, C) $\rightarrow$ T. *GenTag* is the tag generation algorithm. It uses the ciphertext C as inputs and returns tag T for duplicate checking.
- Decry(P, K, C) $\rightarrow$ M. Decry is a symmetric decryption algorithm that takes K, and ciphertext C ($C \in \{0, 1\}^*$) as inputs. M is the decrypted data.

6.1.2   DupLESS and ClearBox

From users' perceptive, CE or MLE suffers from brute-force attack [103] that leads to privacy leakage, particularly for min-entropy (i.e., predicted files). Brute-force attacks are described in detail as follows. If the adversary knows that the target ciphertext C of the target data D is in a specific or known set $S = \{D_1, \ldots, D_n\}$ of size n, the adversary can recover the data D from the set S by off-line encryption. For each $i = 1, \ldots, n$, the adversary simply encrypts D_i via CE to get the ciphertext denoted C_i and returns the D_i such that $C = C_i$, which actually breaks the ciphertext C.

Fig. 6.1 The architecture of
DupLESS or ClearBox

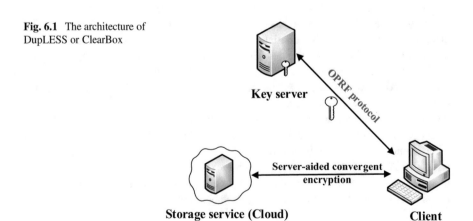

Figure 6.1(a) describes the architecture of DupLESS [103] and ClearBox [107], including client, storage services (Cloud), and key server (KS). To resist brute-force attack, DupLESS [103] and ClearBox [107] employ an oblivious PRF (OPRF) protocol [108] between the KS and clients, which ensures that the KS learns nothing about the client inputs or the resulting PRF outputs, and that clients learn nothing about the key.

DupLESS and ClearBox propose a new server-aided MLE scheme, which com-bines a CE-type base with the OPRF protocol based on RSA blind-signatures and BLS blind-signatures. The BLS-OPRF generates short signatures. The evaluation protocol (EvC, EvS) with verification V_f is presented in [103]. The client users a hash function $H : \{0, 1\}^* \rightarrow Z_N$ to first hash the message to an element of Z_N, and builds the result with a random group element r raised to the eth power. The resulting blinded hash, denoted x, is sent to the KS. The KS signs it by computing $y \leftarrow x^d \bmod N$, and sends back y. Verification then removes the blinding by computing $z \leftarrow y \bullet r^{-1} \bmod N$, then ensures that $z^e \bmod N$ is indeed equal to $H(M)$. Finally, the output of the PRF is computed as $G(z)$, where $G : Z_N \rightarrow \{0, 1\}^k$ is another hash function.

6.2 Privacy Risks and Performance Cost

In Table 6.1, we analyze the problems of state-of-the-art secure deduplication schemes in terms of convergent encryption, key management, and assured deletion. Existing encryption solutions for fine-grained deduplication either suffer from brute-force attacks or incur large time overheads. The key management approaches suffer from large key space overheads and single-point-of-failure. The details are described as follows.

Brute-Force Attacks and Heavy Computation Consumption. CE suffers brute-force attacks because of deterministic and keyless issues [103]. If users' file M is in a known set (e.g., predictable files), the attackers obtain the ciphertexts of target file C. They will traverse all files to generate the ciphertexts and match them with C. To resist brute-force attacks, DupLESS [103] and ClearBox [107], which encrypt data by a message-locked key obtained from a key-server via the oblivious pseudorandom protocol (e.g., RSA-OPRF [103], BLS-OPRF [107]). However, the

Table 6.1 Analysis of secure deduplication schemes in terms of data privacy and key security

Goals	Approaches	Representative work	Limitations
	CE	Farsite [104, 105]	Brute-force attacks
Data privacy	HCE	Pastiche [109, 110]	Brute-force attacks
	DupLESS	DupLESS [103]	Large computation overheads
	Single key server	ClouDedup [111]	Single point of failure
Key security	Master key	DupLESS [103]s	Key space overheads
	Secret splitting	Dekey [112]	Large key space overheads

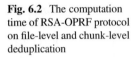

Fig. 6.2 The computation time of RSA-OPRF protocol on file-level and chunk-level deduplication

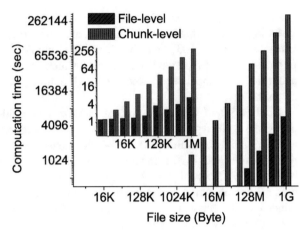

Table 6.2 Data redundancy distribution on four real-world datasets

Datasets (GB)	Total[a] (global)	Cross-user[b] dup files	Inside-user[c] dup chunks	Cross-user[c] dup chunks
One-set	202.1 (100%)	104.28 (51.6%)	78.62 (38.9%)	19.2 (9.5%)
Inc-set	141.3 (100%)	108.8 (77%)	27.13 (19.2%)	11.4 (0.86%)
Full-set	2457.6 (100%)	2393.7 (97.4%)	60 (2.4%)	1.95 (0.02%)
FSLhomes	14463.3 (100%)	13764.7 (95.17%)	687 (4.85%)	11.6 (0.08%)

[a]The size of all redundant data of the datasets
[b]The size of duplicate files from multiple users
[c]The size of duplicate chunks from inter and intra users

OPRF protocol [108] is inefficient and incurs large computation overheads for chunk-level deduplication. Figure 6.2 shows the computation time of RSA-OPRF protocol on file-level and chunk-level deduplication. Note that the latter uses the average chunk size of 4 KB.

Large Key Space Overheads. Users encrypt their data by the convergent keys and protect convergent keys with users' master key [103, 111]. Thus, the number of convergent keys increases linearly with the number of unique data and number of sharing users [112]. Master key suffers from a single-point-of-failure risk [112]. Dekey [112] splits chunk keys into key shares via Ramp Secret Sharing Scheme (RSSS) [113]. As a result, the number of convergent keys explodes with the number of key shares.

To solve the above challenges, it is necessary to propose a new secure deduplication scheme with an efficient key management approach to balance deduplication security and efficiency.

Characteristic	One-set	Inc-set	Full-set	FSL
Num. users	11	6	19	7
Total size (TB)	0.491	0.224	2.5	14.5
Num. files (M)	2.5	0.59	11.3	64.6
Num. chunks (M)	50.5	29.4	417	1703.3
Avg. chunk (KB)	10	8	6.5	8
Dedup factor	1.7	2.7	25	38.6

Table 6.3 Workload characteristics of four real-world datasets

In Table 6.2, we study the data redundancy distribution on the four large real-world datasets (whose workload characteristics are detailed in Table 6.3 in Sect. 6.5). According to the results shown in Table 6.2, we obtain three key observations.

- Cross-user redundant data are mainly from the duplicate files in Table 6.2. Previous works, SAM [76] and Microsoft's study [4], have similar observations.
- Global chunk-level deduplication achieves a high dedup factor, but it brings huge security overheads. Cross-user file-level and inside-user chunk-level deduplication provide security with low overheads, but they achieve a low dedup factor.
- Cross-user and inside-user deduplication schemes face different security challenges, while the latter is lack of potential risks. Specifically, cross-user deduplication requires a secure method with high time overheads, for example, Dup-LESS [103] (e.g., see Fig. 6.2). Inside-user deduplication nevertheless could employ a more efficient method. Motivated by the observations, we employ different secure policies and exploit data semantics for cross-user and inside-user deduplication to balance data security and deduplication performance.

We propose SecDep [148], a semantic-based and low overheads that combines cross-user file-level and inside-user chunk-level secure deduplication to eliminate more redundant data. (1) SecDep employs server-aided HCE at file-level deduplication and user-aided CE at chunk-level deduplication to ensure data security, while significantly reducing security overheads. (2) SecDep uses file-level keys to manage chunk-level keys, which reduces key space overheads.

6.3 Design of SecDep

6.3.1 System Architecture

Our deduplication system consists of a group of clients (e.g., employees) that are connected with an enterprise network. In Fig. 6.3, SecDep is composed of users, storage provider (SP), and distributed key servers (DKS). Users interact with one or more key servers, and backup/restore/delete data to/from the SP. Data are transformed securely by the well-known secure communication protocol (such as SSL/TLS). When a user wants to access the DKS and the SP, the user's passwords

Fig. 6.3 The system
architecture of SecDep,
which contains three entities

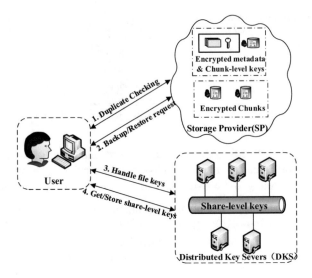

and credentials should be verified. Data stored on the DKS and the SP are protected
by the access control policies.

- Client (user). A client is an entity who wants to back up data to (restore data from)
  the storage provider (SP). The user applies variants of CE to generate keys and
  tags for deduplication, and transfer encrypted chunks to the cloud. The user also
  implements backup/restore/delete protocols.
- Storage provider (SP). The SP maintains tag indices for chunk and file duplicate
  checking. The SP also stores ciphertexts of chunks and chunk-level keys, which is
  detailed in Sect. 6.3.3.
- Distributed key servers (DKS). The DKS is built on a quorum of key servers via
  Shamir Secret Sharing Scheme [114]. Each key server is a stand-alone entity that
  performs the OPRF protocol for convergent key generation [103] and stores
  users' key shares. SecDep can be extended to multiple key servers with threshold
  signatures [115]. To avoid the DoS attack, KS implements rate-liming measures
  [103, 116].

For each file from file streams in SecDep, (1) the user computes a file-level key
and file tag via server-aided hash convergent encryption (HCE). The user sends file
tag to the SP. The SP searches file tag in the file index and returns the checking
results. If the file in not duplicate, the user will divide it into chunks and generate
keys by performing user-aided convergent encryption (CE). The user then sends
the chunk tags to the SP. The SP will check whether the chunk tags exist in the chunk
index and return the results to the user. Finally, the user encrypts all the unique
chunks and sends ciphertexts of chunks to the SP. (2) For key management, the user
encrypts chunk-level keys with file-level key and sends ciphertexts of chunk-level
keys to SP. Then the user splits file-level key into share-level key by the Shamir
Secret Sharing Scheme [114] and sends them to DKS.

6.3.2 UACE: User-Aware Convergent Encryption

In order to resist brute-force attacks and reduce computation overheads, UACE exploits file semantics (e.g., redundant data ownership and data granularity) that combines cross-user file-level and inside-user chunk-level secure deduplication, and adopts different secure policies for better performance.

Cross-user file-level hash convergent encryption (HCE): As shown in Fig. 6.4(a), cross-user file-level hash convergent encryption mainly consists of two steps, i.e., key and tag generations. Specifically, it encrypts data by the server-aided HCE where the CE keys are added secret by a key-server via an oblivious PRF protocol. The oblivious PRF protocol can be built from deterministic blind signatures [103] and we adopt RSA-OPRF scheme, which is based on RSA blind signatures. As long as the server-aided CE key is kept securely, the adversaries cannot break the confidentiality of data. The message sent to DKS is blinded by the random number selected by users themselves. DKS and the adversaries cannot get the file hash even if they obtain the median values.

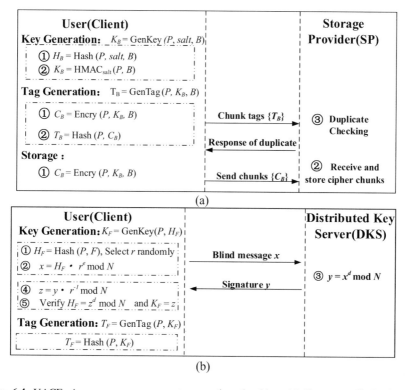

Fig. 6.4 UACE: A user-aware convergent encryption algorithm. (**a**) Cross-user file-level hash convergent encryption. (**b**) Inside-user chunk-level convergent encryption

- GenKey(H_F) is the key generation function. The public RSA exponent e is fixed. The key generation uses e to get (N, d) such that $e \cdot d \equiv 1$ mod $\phi(N)$, where modulus N is the product of two distinct primes of roughly equal length and $N < e$. Then the public key (N, e) and private key (N, d) are returned. For each input file F, the user chooses a random number $r \in N$, gains H_F via computing hash of F and sends $x = H_F \cdot r^e$ mod N mod$N$ to the DKS. The DKS computes y using the equation $y = x^d$ mod N and sends y back. The user just calculates $z = y \cdot r^{-1}$ mod N. The user could also verify whether or not $H_F \equiv Z^d$ mod N. Hence, z is the file-level key.
- GenTag(K_F) is the tag generation function of HCE that takes file key K_F as inputs. File tag could be implemented by $T_F = Hash(K_F)$.

    Inside-user chunk-level convergent encryption: The user encrypts and deduplicates chunks inside-user via user-aided CE. As shown in Fig. 6.4(b), if it is not a duplicate file, UACE will perform inside-user chunk-level deduplication where the key generations are aided by users via adding secret information. The secret information, also called "salt," is kept secure by the user. Then we will introduce key generation, tag generation, and storage of non-duplicate chunks.

- GenKey($salt, B$) is the key generation function that takes a secret information $salt$ and chunk B as inputs. (1) The user calculates chunk hash value$H_B = Hash(B)$. (2) The user computes the chunk-level key via $K_B = HMAV_{salt}(B)$, and HMAC could be implemented by HMAC-SHA256. "Salt" is the secret provided by the user.
- GenTag(K_B, B) is the tag generation algorithm that takes K_B and B as inputs. The user encrypts chunk and gets the chunk ciphertext C_B via $Encry(K_B, B) \rightarrow C_B$. (2) The user gets chunk tag T_B via $Hash(C_B) \rightarrow T_B$. (3) The user sends chunk tag to Storage Provider (SP) for duplicate checking. The SP searches the chunk tag in chunk-tag index and returns the deduplication results to the user.
- Storage of non-duplicate chunks. (1) The user encrypts the unique chunks with their CE chunk-level keys by $Encry(K_B, B) \rightarrow C_B$. (2) The user sends all chunk ciphertexts C_B to the SP. The SP receives C_B and writes them to storage devices.

In general, UACE provides a reasonable trade-off between dedup factor and data security. UACE resists brute-force attacks by employing server-aided HCE at file-level and user-aided CE at chunk-level. Furthermore, UACE proposes fast key generation methods to reduce computation overheads because user-aided CE and HCE offer better performance than sever-aided CE.

6.3.3   MLK: Multi-Level Key Management

Existing convergent key management approaches have variable limitations, single-point-of-failure of keys and large key space overheads [112]. We propose a multi-level key management approach (MLK). In Fig. 6.5, we have a three-level hierarchy,

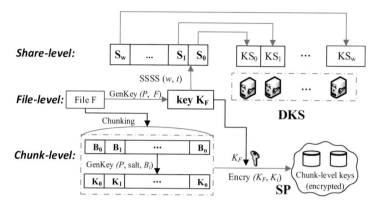

Fig. 6.5 MLK: A multi-level key management approach for secure deduplication, which contains keys at file-level, chunk-level and share-level

namely file-level, chunk-level, and share-level keys. MLK encrypts chunk-level keys with file-level key to reduce key space overheads. To avoid single-point-of-failure, MLK splits file-level key into share-level via Shamir Secret Sharing Scheme and sends secure share-level keys to DKS.

- File-level: For each input file F, MLK generates the file-level key K_F by *GenKey*(H_F).
- Chunk-level: For a non-duplicate file F, it will be divided into chunks $\{B_i\}(i = 1, 2, ...)$. Then, each chunk B_i is assigned a key K_i by *GenKey*(K_F , K_i). MLK encrypts $\{K_i\}$ with K_F by *Encry*(K_F, K_i). The ciphertexts of chunk-level keys will be sent to the SP.
- Share-level: As for file key K_F , MLK splits it into secure share-level keys $\{S_j\}($ $j = 1, 2, ..., w)$ by Shamir Secret Sharing Scheme (w, t). And MLK distributes $\{S_j\}$ to distributed key servers (DKS) through secure channel. Each key server stores and ensures security of share-level keys.

6.4   Security Analysis

SecDep is designed to ensure data confidentiality and key security for cross-user fine-grained deduplicated system for cloud backups. In SecDep, we consider two types of adversaries, that is, external adversary and internal adversary. However, we only focus on internal adversary because SecDep could resist the external attacks by authentication [112]. We assume that the following technologies are secure, such as Shamir Secret Sharing Scheme [114] and symmetric encryption. We analyze data confidentiality and security of keys in the case the adversary compromises the SP, colludes with users, or steals data from key servers. Thus we present security analysis of SecDep in a multi-tiered way.

6.4.1   Confidentiality of Data

In the case that the adversary tries to compromise the SP or collude with users, SecDep could resist brute-force attacks and duplicate-faking attacks to ensure data security, including the confidentiality and integrity. The adversary tries to obtain the content of files from other legitimate users. The adversary may compromise the SP to get the chunks on the SP and perform brute-force attacks. Specifically, the adversary obtains the ciphertexts of target chunks from a specific file. The adversary knows that the chunks are from a specific set $|S|$. For each chunk, the adversary encrypts it to get the ciphertext and compares it with the target chunk. Then the adversary gets the original file. However, SecDep can still ensure data confidentiality. All users' data that are uploaded to the SP have been encrypted with chunk-level keys. The chunk-level keys are generated by adding secret via user-aided CE.

In general, it is difficult to break the confidentiality of users' data because the adversary does not know the secret. The adversary colludes with some users and performs duplicate-faking attacks [105, 110] to the data on the SP, which compromises the integrity of users' data. Specifically, the adversary and these colluders may upload their data to the deduplicated system for cloud backups. They upload the correct tags, but replace the chunks with the wrong data. To address this problem, SecDep can compute the hash value of the ciphertext of chunks. Then the user compares hash value with its tag to resist duplicate-faking attacks.

6.4.2   Security of Keys and SecDep

The adversary tries to obtain the keys, and recover other users' data. Specifically, (1) the adversary gets chunk-level keys by compromising the SP. However, chunk-level keys are encrypted by file-level key via symmetric encryption. SecDep can ensure the security of chunk-level keys as long as file-level keys are stored securely. (2) The adversary tries to get share-level keys from key servers and recover the file-level key, which is very difficult. This is because the share-level keys in SecDep are securely distributed in several key servers by using the known Shamir Secret Sharing Scheme (SSSS) [114].

As mentioned above, the adversary cannot obtain other users' data if it only compromises the SP, colludes with users or steals data from key servers. Then we discuss the security of SecDep if the adversary tries to compromise the SP or collude with users.

In the best case, the adversary compromises the SP, but cannot access to the DKS. All data and metadata stored in the SP is encrypted with random keys, including file recipe. The adversary cannot know the content of other users' data even if it performs brute-force attacks. In the semi-best case, the adversary has compromised some users and have been authorized access to DKS. SecDep can still ensure data security. Although adversary can perform brute-force attacks on chunks via CE, he/she cannot break the encryption key due to not knowing the user's secret. In the worst case, if the adversary has obtained some users' secret and other users' ciphertexts of chunks, SecDep can still ensure security for unpredictable data that are not falling into a known set. The worst case is that adversary attacks the predictable data within a known set. SecDep makes this worst case rarely occur by further protecting tags and file metadata [103].

6.5 Performance Evaluation

6.5.1 Experiment Setup

In order to evaluate the performance of SecDep, we conduct the experiments using machines equipped with an Intel(R) Xeon(R) E5606@2.13GHZ 8 Core CPU, 16GB RAM, and installed with Ubuntu 12.04 LTS 64-bit Operation System. These machines are connected with 100 Mbps Ethernet network. We implement a research prototype to evaluate and compare the performance of different schemes, including Baseline, DupLESS-file, DupLESS-chunk, and SecDep. Baseline is a basic chunk-level deduplication system without any security mechanisms. DupLESS-file and DupLESS-chunk are implementing secure DupLESS [103] schemes at file and chunk levels respectively. Note that our evaluation platform is not a production quality secure deduplication system but rather a research prototype. Hence, our evaluation results should be interpreted as an approximate and comparative assessment of other secure deduplication approaches above, and not be used for absolute comparisons with other deduplication systems.

We use both of the synthetic datasets and real-world datasets for evaluation. The synthetic datasets consist of artificial files filled with random contents and each file is divided into fixed-sized chunks. Real-world datasets are summarized in Table 6.3, including One-set, Inc-set, Syn-set, Full-set, and FSL (FSLhomes). One-set was collected from 11 graduate students of a research group and was reported by Xia et al. [117]. Inc-set was collected from initial full backups and subsequent incremental backups of 6 members of a university research group and was reported by Tan et al. [76]. Syn-set contains 200 synthetic backups and the backup is simulated by the file create/delete/modify operations [118]. Full-set consists of 380 full backups of 19 researchers' PC and is reported by Xing et al. [78]. FSLhomes is a public dataset reported by Tarasov et al. [118] and can be downloaded from the website [119]. FSLhomes contains snapshots of students' home directories, where files consist of source code, binaries, office documents, and virtual machine images.

In order to compare the performance of the existing secure deduplication schemes and SecDep, we mainly use dedup factor, backup time, key space, key protecting time, and deletion time as the quantitative metrics. Dedup factor is defined as the ratio of the data sizes before/after deduplication. Backup time consists of key generations, tag generations, searching, encryption, and transferring time. Key space is the key storage overheads. Key protecting time is the encryption or encoding time for key management. We observe the impacts of varying sizes and numbers of files and average chunk sizes on the system performance.

6.5.2   Deduplication Ratio and Backup Time

In Fig. 6.6, we evaluate the relative dedup factor of the five real-world datasets. The Y-axis shows the relative dedup factor to DupLESS-chunk. The normalized dedup factor of DupLESS-chunk is regarded as 100%, while the real dedup factor is listed in Table 6.3. We evaluate the relative backup time of DupLESS-file, SecDep, and DupLESS-chunk under these datasets. The backup time of DupLESS-chunk is regarded as 100%. We measure and record the average time of each individual step, for example, tag and key generations at file-level and chunk-level, index searching, encryption, and data transfer. Based on these times, we obtain the backup time of four real-world datasets.

As shown in Fig. 6.6, SecDep eliminates the majority of duplicate data, only resulting in a 2.8–7.35% loss of dedup factor compared with the DupLESS-chunk as discussed in Sect. 6.2. It is because that SecDep combines cross-user file-level and inside-user chunk-level deduplication, which efficiently eliminates most of duplicate data in backup datasets.

Figure 6.7 shows the time overheads of DupLESS-File, SecDep, and DupLESS-Chunk on real-world datasets. The Y-axis shows the relative backup time to DupLESS-chunk. It suggests that SecDep reduces 52–92% of backup time overheads com-pared with DupLESS-chunk. There are two reasons: (i) When files are duplicates, SecDep just performs file-level deduplication via server-aided HCE, which offers better performance than DupLESS-file using the conventional CE (ii) SecDep uses user-aided CE inside-users to generate keys, which avoids the time-consuming RSA-OPRF operations of DupLESS-chunk. Figure 6.7 shows that

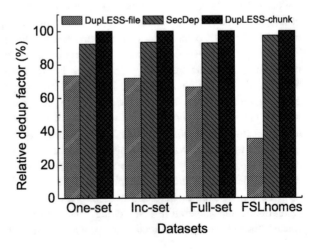

Fig. 6.6 The comparisons of DupLESS-file, SecDep, and DupLESS-chunk in terms of relative dedup factor

Fig. 6.7 The time
overheads of DupLESS-file,
SecDep, and
DupLESSchunk on real-
world datasets

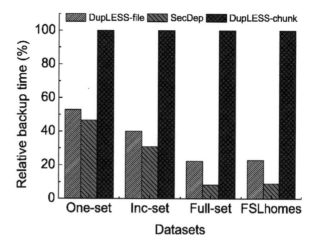

Table 6.4 The key space
overhead comparison among
different chunk-level key
management approaches

Approach	Key space overheads
Single key server	$O = \frac{D^*S}{k^*DF}$
Master key	$O = \frac{D^*S}{k}$
SecDep	$O = \frac{D^*S}{k^*DF} + N_u{}^*w^*S$
Dekey	$O = \frac{D^*S}{k^*DF} * \frac{w}{t-r}$

SecDep is more time-efficient than DupLESS-file. When files are duplicate, SecDep reduces much more time overheads on key and tag generations compared with DupLESS-file because HCE offers better system throughput than CE. SecDep reduces more transfer time compared with DupLESS-file due to the higher dedup factor.

6.5.3 Space and Computation Overheads

We evaluate the space and computation overheads of SecDep's multi-level key management. First, we describe the key space overheads (O) of different approaches in Table 6.4. The parameters used in Table 6.4 are defined as follows. D is data size, N_u is the number of unique file, k is the average chunk size, S is the size of chunk key, and DF is the dedup factor. RSSS(w, t, r) [112] and SSSS(w, t) [114] take w, t, and r as parameters, which are set to 6, 4, and 2, respectively.

In Table 6.4, the key space of Dekey is $w/(t - r)$ times the number of chunk-level keys [112]. SecDep adds $N_u *w *S$ shares over the single key server approach, which only accounts a very small fraction of the total key space overhead. In Fig. 6.8, the key space overheads of SecDep will be decreased with the deduplication factor and SecDep adds little key space overheads compared to the single key server approach. Note that the master key approach has the highest key space overheads. This is

Fig. 6.8 The key space overhead per file of different schemes under various dedup factors

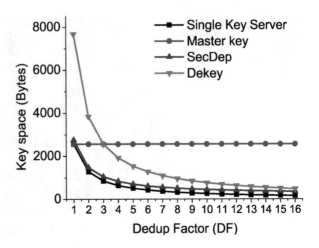

Fig. 6.9 The key space overheads consumed by the four key management approaches on the four real-world datasets

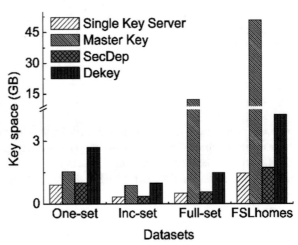

because the Master Key approach always encrypts chunk-level keys with users' own master key to check whether the files are duplicates or not.

Figure 6.9 suggests that SecDep reduces 59.5–63.6% and 34.8–96.6% of key space overheads on the five real-world datasets compared with Dekey and Master Key approach, respectively. SecDep uses file-level keys to manage chunk-level keys to avoid the key space overheads increasing too fast as discussed in the Sect. 6.3.3.

6.6   Concluding Remarks

This chapter describes the contradiction between data encryption and deduplication, and the complexity of key management. The current algorithm convergent encryption or DupLESS fails to solve the problems due to brute-force attacks, large key space overheads, and system performance. This chapter presents SecDep, which employs variants of convergent encryption on inter- or intra-user by exploiting redundant data distribution, and leverages multi-level key management by secret sharing. As a result, SecDep ensures data and key privacy and reduces computation and key space overheads. Evaluation results based on real-world datasets demonstrate that SecDep achieves security goals and cost savings.

Chapter 7
Post-deduplication Delta Compression Schemes

Abstract Delta compression has been gaining increasing attention in recent years for its ability to remove redundancy among non-duplicate but very similar data files and chunks, for which the data deduplication technology often fails to identify and eliminate. Given a new chunk B and an existing chunk A, delta compression encodes B relative to A and generates their differences. We call the differences a "delta" and the chunk A its "base." We then only need to transfer or store the delta, rather than the entire chunk B, thus obtaining bandwidth- or space-savings. Due to significant data reduction efficiency, some applications adopt delta compression as a complement for chunk-level deduplication to further reduce the storage space (or bandwidth) requirement. In this chapter, we discuss the problems facing the post-deduplication delta compression and the solutions to address the problems. The rest of this chapter is organized as follows: Section 7.1 presents the state-of-the-art solutions for post-deduplication delta compression techniques. Section 7.2 describes the design and implementations of our proposed deduplication-inspired delta compression approach, called Ddelta. Section 7.3 describes our proposed delta compressed and deduplicated technique with low overhead.

7.1 Post-Deduplication Delta Compression Techniques

The Super-Feature Approach. Delta compression is costly in the sense of requiring extra computation and I/O. To achieve a good trade-off between the benefits and the cost, it is expected that two chunks for delta compression are highly similar. The process of detecting a fairly similar candidate for delta compression is called resemblance detection. The most widely used resemblance detection approach is super-feature approach [120] which is based on Broder's theorem.

Recall that Broder's theorem states that the probability of two sets H(S1) and H(S2) having the same minimum hash element is the same as their Jaccard similarity coefficient [121], where S1 and S2 are two sets and H(S1) and H(S2) are the corresponding sets of the hashes of the elements of S1 and S2 respectively. In other words, the minimum elements of H(S1) and H(S2) can be selected as features to measure the similarity degree of S1 and S2.

© Springer Nature Singapore Pte Ltd. 2022
D. Feng, *Data Deduplication for High Performance Storage System*,
https://doi.org/10.1007/978-981-19-0112-6_7

Based on this theorem, the super-feature approach detects the similarity for delta compression by deterministically sampling several Rabin fingerprints as features and coalescing them into a super-feature, also referred to as super-fingerprint. Specifically, the approach uses a rolling hash function of fixed-size (e.g., 48 bytes) windows over all overlapping small regions of data, and chooses the minimal hash value as the feature. By using multiple different hash functions, it generates multiple features. In practice, to generate multiple, independent features, super-feature first generates a Rabin fingerprint over rolling window w of a file/chunk, and then permutes the Rabin fingerprint to generate multiple values with randomly generated coprime multiplier.

SIDC. EMC proposed SIDC [7], a stream-informed delta compression approach to reduce the WAN bandwidth requirement. SIDC shows that index for resemblance detection can overflow the RAM capacity, and stream-informed locality is effective for delta compression. Specifically, remote repository of SIDC caches a small part of sketches of the chunks for resemblance detection. To reduce the volume of data transferred, delta compression should be performed before sending to the remote repository. Thus, after deduplication, backup server sends the sketches of non-duplicate chunks to the remote repository. Then the remote repository searches the similar chunks (base chunks) for the non-duplicate chunks by referring to the sketch cache, sends the fingerprints of detected similar chunks to the backup server. Finally, the backup server checks whether the base chunks exist in local. If yes, it performs delta compression and sends the delta to the remote repository, thus obtaining a bandwidth saving. Otherwise, it sends the chunks to the remote repository. Because of stream-informed locality, SIDC attains significant benefits from delta compression.

Difference Engine. Gupta et al. [49] proposed Difference Engine (DE), an extension to the Xen virtual machine monitor, to save memory usage in VM environments. The main idea behind DE is exploiting page similarity and performing delta compression for similar pages to decrease the memory usage. In detail, assuming that there are two VMs. For the identical pages across the two VMs, DE stores a single copy and creates references that point to the original page. For pages that are similar but not identical, DE delta compresses other page using Xdelta [122] and stores only the patch against the reference page. For pages that are unique and infrequently accessed, DE directly compresses the pages (e.g., with LZO algorithm) in memory. Because there are quite a percentage of similar pages among VMs in a VM environment, DE gains the benefit of delta compression and hence effectively enlarges the logical memory space.

Other Delta Compression Schemes. REBL [123] and DERD [124] are typical super-feature-based resemblance detection approaches for data reduction. They compute the features of the data stream (e.g., Rabin fingerprints [61]) and group features into super-features to capture the resemblance of data and then delta compress the data. TAPER [125] proposes an alternative to the super-feature method by representing each file with a Bloom filter that records chunk fingerprints, measuring file similarity based on the number of matching bits between Bloom filters, and then delta compressing the detected similar files. I-CASH [126] utilizes delta compression to remove similar data so as to save space and enlarge the logical space of SSD caches, where it detects similar 4KB pages using a parameterized scheme based upon computing the hashes of several 64B subpages, which is similar to the super-features approach.

Challenges of Delta Compression. In general, one of the main challenges facing the post-deduplication delta compression technology is the high computation overheads of delta encoding. State-of-the-art delta compression techniques, such as Xdelta [122] and Zdelta [127], use a method similar to that of the traditional lossless compression approaches, namely, a byte-wise sliding window to identify matched (i.e., duplicate) strings between the base chunk and the target chunk for the delta calculation, which is very time-consuming.

Another challenge is the computation and memory overheads required by resemblance detection. To index a dataset of 80 TB and assuming an average chunk size of 8 KB and 16 bytes per index entry, for example, about 200 GB worth of super-feature index entries must be generated, which is too large to fit in RAM [7]. Moreover, since the random accesses to on-disk index are much slower than that to RAM, the frequent accesses to on-disk super-features will cause the system throughput to become unacceptably low for the users [7, 16, 73].

To address the two challenges above, we propose Ddelta [34] and DARE [33]. Ddelta is a deduplication-inspired fast delta compression scheme that effectively leverages the simplicity and efficiency of data deduplication techniques to improve delta encoding/decoding performance. And DARE is a low-overhead deduplication-aware resemblance detection and elimination scheme for backup and archiving storage system.

7.2 Ddelta: A Deduplication-Inspired Fast Delta Compression

This section describes Ddelta, a deduplication-inspired fast delta compression scheme that effectively leverages the simplicity and efficiency of data deduplication techniques to improve delta encoding/decoding performance.

7.2.1 Computation Overheads of Delta Compression

Deduplication [16, 18], as an approach to data reduction in storage system, divides the data stream into independent chunks of approximately equal length by CDC (e.g., Rabin-based Chunking) algorithm [128] and then uses secure fingerprints (e.g., SHA-1) to uniquely identify these chunks. If any two secure fingerprints match, deduplication will consider their corresponding chunks also identical. Thus the deduplication approach simplifies the process of duplicate detection and scales well in mass storage systems, especially in backup/archival systems with abundant duplicate contents [6]. But deduplication only identifies duplicate chunks/files and thus fails to detect the non-duplicate but very similar chunks/files, which can be supplemented by the delta compression technology.

While delta compression achieves a superior performance of data reduction among similar data chunks, the challenge of the time-consuming string matching stemming from delta encoding remains [32, 49, 126]. The study of Difference Engine shows that delta compression (i.e., page sharing) consumes a large amount of CPU time due to the overheads of resemblance detection and delta encoding [49]. Shilane et al. [32] suggest that the average delta encoding speed of similar chunks falls in the range of 30–90 MB/s, which may become a potential bottleneck in storage systems. This is because delta compression finds the matched strings by using a byte-wise sliding window. If the string of the sliding window does not find a match, the sliding window will move forward by only one or several bytes at a time [122].

Inspired by deduplication that divides the data streams into independent and nonoverlapping chunks to simplify the duplicate detection, we believe that the string matching process can also be simplified by the chunking concept of deduplication. This means that delta compression does NOT need to employ the byte-wise sliding window to find the matching strings. Instead, we divide the base and input chunks into smaller, independent, and nonoverlapping strings by CDC and identify duplicates only among strings. Ddelta is motivated by deduplication to employ a simple content-based sliding window to find a chunking boundary to generate nonoverlapping strings for duplicate identification in delta compression. As a result, Ddelta is able to generate fewer strings than Xdelta, which accelerates the process of string matching for delta calculation.

However, the CDC-based approach may generate non-duplicate but very similar strings (e.g., two very similar strings with only a few bytes being different) and lead to failure in identifying redundancy among them. We will address this problem of compromised compression ratio due to the possible inaccurate boundary identification of CDC-based chunking in Sect. 7.2.3 where we take inspirations from studies on content locality of redundant data [16, 67, 117] to search for likely more redundancy in the duplicate-adjacent areas/regions.

7.2.2   Deduplication-Inspired Delta Compression

The design goal of Ddelta is to accelerate the duplicate detection process in the encoding phase of delta compression for similar data files and chunks. Algorithm 1 describes the key workflow of deduplication-inspired Ddelta encoding that is designed to speedup encoding. For data chunks that are either identified to be similar via resemblance detection [32, 123, 129] or already known delta compression candidates [122, 130, 131], Ddelta is applied to fast delta encoding these data chunks as follows:

Algorithm 1: Deduplication-Inspired Ddelta Encoding

Input: base chunk, *src*; input chunk, *tgt*
Output: delta chunk, *dlt*;
1: **function** COMPUTDELTA(*src, tgt*)
2:    *dlt* ←*empty*
3:    *Slink*← GEARCHUNKING(*src*)
4:    *Tlink*← GEARCHUNKING(*tgt*)
5:    *Sindex*← INITMATCH(*Slink*)
6:    *str*←*Tlink*; *len*←*size*(*str*)
7:    **while***str*! = *NULL* do
8:        *pos*← FINDMATCH(*src,Sindex, str, len*)
9:        **if** *pos* <0 **then**   ◄No matched string
10:            *dlt*+ =Instruction(*Insert str, len*)   ◄Add the unmatched string into the delta chunk
11:        **else**
            ◄Find a matched string
12:            *dlt*+ =Instruction(*Copy pos, len*)   ◄Add the matched info into the delta chunk
13:        **end if**
14:        *str*←(*str*→*next*); *len*←*size*(*str*)
15:    **end while**
16:    **return** *dlt*
17: **end function**
18:
19: **function** INITMATCH(*Slink*)
20:    *str*←*Slink*; *pos*←0
21:    *Sindex*←*empty*
22:    **while***str*! = *empty* **do**
23:        *f* ←Spooky(*str, size*(*str*))
24:        *Sindex*[*hash*[*f*]]← *pos*
25:        *pos*+ = *size*(*str*)
26:        *str*←(*str*→*next*)
27:    **end while**
28:    **return** *Sindex*
29: **end function**
30:
31: **function**FINDMATCH(*src,Sindex, str, len*)
32:    *f* ←Spooky(*str, len*)
33:    **if***Sindex*[*hash*[*f*]] = *empty***then**
34:        **return**−1
      ◄No matched fingerprint
35:    **end if**
36:    *pos*←*Sindex*[*hash*[*f*]]
37:    **if** memcmp(*src*+ *pos, str, len*)= 0 **then**
38:        **return***pos*
        ◄Find a matched string
39:    **else**
40:        **return**−1
      ◄Spooky hash collision
41:    **end if**
42: **end function**

Gear-based chunking. Ddelta divides the similar chunks into independent and nonoverlapping strings based on the Gear hashes of their content, which is implemented by the GEARCHUNKING() function in Algorithm 2.

Spooky-based fingerprinting and duplicate identification. Ddelta identifies the duplicate strings obtained from Gear-based chunking above by computing and then matching their Spooky hashes. If their Spooky hashes match, their content will be further verified for duplicates, as implemented in the FINDMATCH() function of Algorithm 1.

Greedily scanning duplicate-adjacent areas. To improve the duplicate detection and thus detect more duplicates, Ddelta byte-wisely scans non-duplicate areas immediately adjacent to the above-confirmed duplicate strings exploiting content locality of redundant data.

Encoding. Ddelta encodes the duplicate and non-duplicate strings as "Copy" and "Insert" instructions respectively.

Algorithm 2: Gear-Based Chunking for Ddelta Encoding

> **Input**: chunk, *src*;
> **Output**: strings linklist, *Slink*;
> 1: Predefined values: mask bits for the average string size, *mask*; matched value, *xxx*
> 2: **function** GEARCHUNKING(*src*)
> 3: $fp \leftarrow 0$; $pos \leftarrow 0$; $last \leftarrow 0$
> 4: $SLink \leftarrow empty$
> 5:   **while** $pos < size(src)$ **do**
> 6: $fp = (fp \ll 1) + GearTable[src[pos]]$
> 7:   **if** $fp \ \& mask = xxx$ **then**
>   ◄Get the MSB of fp
> 8: $str = src[last \rightarrow pos]$
> 9:   InsertLinkList($str, Slink$)
> 10: $last \leftarrow pos$
> 11:   **end if**
> 12: $pos \leftarrow pos + 1$
> 13:   **end while**
> 14:   **return** $SLink$
> 15: **end function**

7.2.3 Gear-Based Fast Chunking

We propose a fast hash-based chunking algorithm, called Gear-based chunking, to accelerate the CDC scheme for Ddelta by using the re-hashed ASCII values of data content of similar chunks, as shown in Fig. 7.1. In Gear-based chunking, the hash value of the sliding window is cumulatively and quickly calculated from the previous value by the GearHash in the form of "$Hi = (Hi-1 \ll 1) + GearTable[Bi]$".

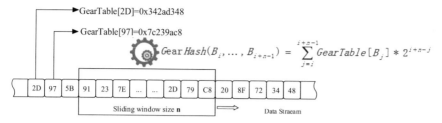

Fig. 7.1 Gear-based fast chunking

Algorithm 2 describes the workflow of Gear-based chunking for delta encoding. Gear-based chunking determines the chunking boundary by checking whether the value of the x most-significant bits (MSB) of Gear hash is equal to a predefined value or not. Note that this MSB-based way of finding the boundary by Gear-based chunking is different from Rabin-based chunking in that the latter determines the chunking boundary by checking if the Rabin hash value mod M, i.e., the value of the x least-significant bits (LSB), is equal to a predefined value, where M is the average chunk size. Note that the number $\times$ of MSB or LSB bits is determined by the average chunks size M as $x = \log2(M)$. For example, to generate chunks of average 8 KB size, Gear-based chunking would choose the 13 most-significant bits (i.e., log2 (8192)) of the Gear hash value to determine the chunking boundary, whereas Rabin-based chunking would choose the 13 least-significant bits.

7.2.4   Greedily Scanning the Duplicate-Adjacent Areas

The CDC-based approach [18, 128] cannot always accurately find the boundary between the changed and duplicate regions among similar data and, instead, simply determines the chunk boundary if the hash (e.g., Rabin, Gear) of the CDC sliding window matches a predefined value. Consequently, for some non-duplicate but nearly identical strings (i.e., with only one or two bytes being different), Ddelta may generate totally different Spooky fingerprints and thus miss the opportunity to identify redundancy among these strings. To address this challenge, we take inspirations from the studies of Bimodal Chunking [67] and SiLo [117] on the content locality of redundant data, which suggests that the neighboring data of duplicate chunks should be considered good deduplication candidates due to the data-stream content locality. Therefore, Ddelta also searches the areas immediately adjacent to known duplicates to lend themselves to easy duplicate detection by the following two steps.

- Chunk-level search in the duplicate-adjacent areas. For those resemblance-detected chunks [32, 49, 123] (i.e., the delta compression candidates), Ddelta will directly search and scan from both ends of similar chunks toward centers for duplicate matching until byte-wise comparison fails to find a match. For example, given contiguous chunks {A, B, C} and {A', B', C'} from two data streams where A' and C' are detected to be duplicate to A and C respectively and B' is resemblance-detected to be very similar to B, Ddelta can search the duplicate-adjacent areas, namely, the beginnings and the endings of B' and B, to detect more duplicate contents due to the content locality of data stream.

- String-level search in the duplicate-adjacent areas. For those duplicate-detected strings, Ddelta byte-wisely searches and scans their immediately adjacent non-duplicate strings to identify more duplicates in a way similar to above chunk-level search.

7.2.5   Encoding and Decoding

Combining the techniques of deduplication-inspired delta encoding and greedily scanning for more duplicates, Ddelta encodes the matched and unmatched regions of the input chunk as a delta chunk by the two instructions of "Copy[offset,length]" and "Insert[data,length]" respectively. For those matched or unmatched contiguous data regions, Ddelta will merge them into a single "Copy" or "Insert" instruction to simplify both the delta encoding and decoding operations.

To restore the input chunk, Ddelta decodes the instructions in the delta chunk sequentially. For a "Copy" instruction, Ddelta reads the data from the base chunk according to the information of offset and length. For an "Insert" instruction, Ddelta directly reassembles the data from the delta chunk into the restored chunk.

7.2.6   The Detail of Ddelta Scheme

To put things together and in perspective, Fig. 7.2 shows the Ddelta encoding workflow by way of an example. For each pair of resemblance-detected similar chunks or already known delta compression candidates, Ddelta goes through the following four key steps.

- Step 1. Chunk-level greedily scanning for duplicates as illustrated in Fig. 7.2(a).
- Step 2. Duplicate-string identification by Gear-based chunking and Spooky-based fingerprinting as illustrated in Fig. 7.2(b).
- Step 3. String-level greedily scanning for duplicates as illustrated in Fig. 7.2(c).
- Step 4. Generate delta chunk based on results of duplicate identification from steps 1–3 as illustrated in Fig. 7.2(d).

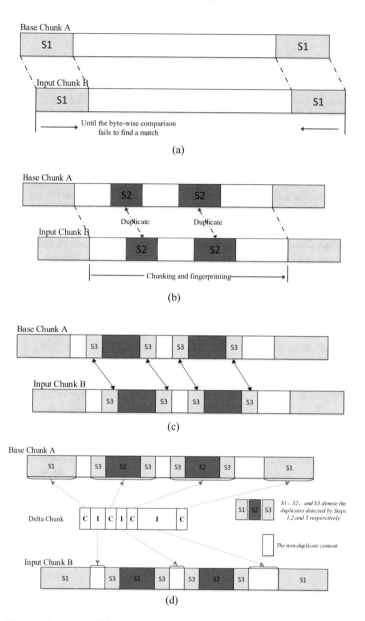

Fig. 7.2 The workflow of the Ddelta approach. (**a**) Step1: Scanning from both ends of two known similar chunks (i.e., A and B that are adjacent to confirmed duplicate chunks (not shown here)) toward the center in search of duplicate content by byte-wise comparison, until match fails. (**b**) Step 2: Using Gear-based chunking to divide the remaining areas (i.e., non-duplicate portions of A and B after Step1) into smaller and nonoverlapping strings and identifying duplicate strings by their Spooky fingerprints and content. (**c**) Step 3: Scanning the areas immediately adjacent to duplicate strings identified in Step 2 for additional duplicate content. (**d**) Step 4: Encoding duplicate and non-duplicate data regions into the delta chunk, where C and I stand for the "Copy" and "Insert" instructions respectively

7.2.7   Performance Evaluation

Experimental Platform. We implement and evaluate a Ddelta prototype on the Ubuntu 12.04.2 operating system running on a quad-core Intel i7 processor at 2.8 GHz, with a 16 GB RAM, two 1 TB 7200RPM hard disks, and a 120 GB SSD of Kingston SVP200S37A120G.

Configurations for Data Reduction. For data deduplication, the average, maximum, and minimum chunk sizes are 8 KB, 64 KB, and 2KB respectively, which is the same configuration as the one in LBFS [18]. For delta compression, two well-known open-source projects, Xdelta [122] and Zdelta [127], are used in the evaluation as the baselines for the proposed Ddelta compression. In addition, we use the GZIP compression [132] (short for GZ) to assess the post-delta-compression data reduction performance.

Evaluation Datasets. Six datasets are used in the evaluation of Ddelta, with their key characteristics summarized in Table 7.1 and detailed as follows.

- GCC and Linux are two well-known open-source projects that are available from the websites [77, 133]. These two datasets represent workloads of typical large software source code.
- VM-A is a VM archive that consists of 78 virtual machine images of different OS release versions from the website [134], including 23 Centos images, 18 Fedora images, 17 Ubuntu images, 12 FreeBSD images, and 6 Debian images. This dataset represents the workload with low deduplication factor.
- VM-B is a VM backup dataset containing 177 full backups of an Ubuntu 12.04 virtual machine in use, which is a common use-case for data reduction in the real world.
- RDB is a dataset collected from the Redis key-value store database [135]. We backup the dump.rdb files as the snapshots of the Redis database, and collect 200 backups, which represents a typical database workload for data reduction.
- Bench is a benchmark dataset generated from the snapshots of a personal cloud storage benchmark [130]. We simulate common operations of file systems, such as file create/delete/modify on the snapshots of this benchmark according to existing approaches [83, 118].

Table 7.1 Workload characteristics of the six datasets used in the performance evaluation. Here deduplication factors measured by Rabin-based chunking

Dataset name	# of Images/versions	Total size	Deduplication factor
GCC	43	14.1 GB	6.71
Linux	258	104 GB	44.7
VM-A	78	114 GB	1.65
VM-B	117	1.78 TB	25.8
RDB	211	1.15 TB	7.22
Bench	200	1.54 TB	35.0

Fig. 7.3 The workflow of a post-deduplication data reduction (delta + GZ compression) system

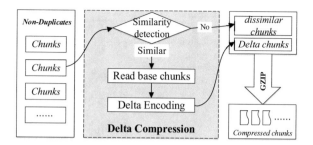

Table 7.2 Comparisons of compression ratio, compression factor, and the real average chunk size between Rabin-based chunking and Gear-based chunking for data deduplication with the expected average chunk size of 8 KB

Dataset		GCC	Linux	VM-A	VM-B	RDB	Bench
Compression ratio (CR)	Rabin%	85.09	97.76	39.50	96.13	86.15	97.14
	Gear%	85.09	97.76	39.83	96.15	86.18	97.15
	Difference%	0.003	0.001	0.329	0.035	0.023	0.009
Compression factor (CR)	Rabin	6.71	44.69	1.65	25.84	7.22	35.03
	Gear	6.71	44.71	1.66	25.99	7.24	35.13
	Difference	0.001	0.015	0.009	0.156	0.018	0.106
Average chunk size	Rabin	4.19	5.80	12.4	10.9	9.55	9.95
	Gear	4.20	5.83	11.3	10.5	9.54	9.93
	Difference	0.011	0.033	1.144	0.395	0.008	0.019

Table 7.3 Compression factors of deduplication and post-deduplication data reduction

Dataset	Compression factor (CF)				Final
	Dedupe	Delta	GZ	Total	Size
GCC	6.71	2.58	2.90	50.2	287 MB
Linux	44.7	3.14	3.09	434.5	245 MB
VM-A	1.65	1.60	2.51	6.63	17.2 GB
VM-B	25.8	1.55	2.35	93.7	19.0 GB
RDB	7.22	5.29	1.47	56.3	20.4 GB
Bench	35.5	2.23	1.00	78.4	19.6 GB

Evaluation of Ddelta. We evaluate the compression efficiency and delta encoding/decoding speed of Ddelta by means of a post-deduplication data reduction system that implements delta and GZ compression on the non-duplicate chunks (i.e., data deduplication is already done) as shown in Fig. 7.3. The results are shown in Tables 7.2 and 7.3. Delta compression achieves a CF of more than 3 on the Linux and RDB datasets and CF of $1.5\times$–$2.5\times$ on the other four datasets. GZ compression achieves a CF of more than $2\times$ on the first four datasets but does relatively poorly on the last two datasets that contain a large amount of random-byte content.

Figure 7.4(a) shows a comparison in encoding speed between the deduplication (Rabin + SHA1) and Ddelta (Gear + Spooky) solutions as a function of the chunked string size. Ddelta accelerates the deduplication-based delta encoding by a factor of

Fig. 7.4 Comparisons in delta encoding/decoding speed among deduplication, Ddelta, Xdelta, and Zdelta. (**a**) Ddelta vs. deduplication. (**b**) Encoding speed. (**c**) Decoding speed

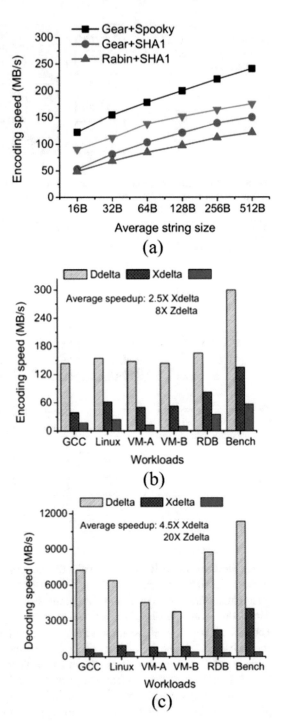

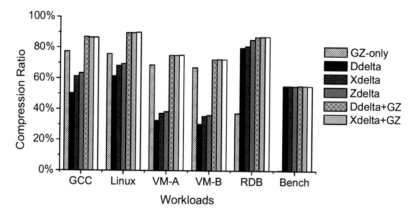

Fig. 7.5 Comparison in *CR* (compression ratio) among seven post-deduplication data reduction schemes

about 1.7× via Spooky-based fingerprinting and by a factor of about 1.4× via Gear-based chunking. This indicates that the Gear-based and Spooky-based approaches contribute significantly to Ddelta's ability to reduce the compute overhead for delta encoding. Figure 7.4(b) and (c) compare the delta compression performance of Xdela, Zdelta, and Ddelta (with an average string size of 32B) on the six datasets. Ddelta is shown in Fig. 7.4(b) to achieve an average delta encoding speedup of 2.5× and 8× over Xdelta and Zdelta respectively, by chunking the similar data into independent and nonoverlapping strings to simplify the duplicate detection. Meanwhile, Ddelta also achieves a decoding speedup of 4.5×–20× over Xdelta and Zdelta by using Gear-based chunking to generate nonoverlapping strings for duplicate identification. Note that Zdelta has the lowest encoding and decoding speeds because of its addition of the Huffman coding for delta compression, which improves redundancy elimination slightly but at the expense of longer processing time.

Figure 7.5 shows the CR of different combinations of various data reduction approaches. Ddelta (with an average string size of 32B) achieves nearly 90% of compression ratio of Xdelta and Zdelta when the post-deduplication data reduction system only applies delta compression on top of deduplication. But when the system combines Delta and GZ compression together for data reduction, the Ddelta+GZ ap-proach achieves nearly the sameCR as the Xdelta+GZ and Zdelta+GZ solutions. GZ compression achieves better data reduction performance than delta-compression-only approaches on the first four datasets, but their combined GZ +Delta solution obtains a superior performance of CR to the GZ-only approach. Meanwhile, GZ compression performs relatively poorly on reducing the last two datasets that contain significant random-byte content (CR of GZ-only approach on the Bench dataset is nearly equal to zero), which were already well compressed by delta compression solutions.

Based on the results from Figs. 7.4 and 7.5, Ddelta achieves a superior perfor-mance of delta encoding and decoding but at a cost of slightly lower compression ratio, which can be supplemented by GZ compression as shown in Fig. 7.5. The combined Delta and GZ compression achieves a superior performance of data reduction but at the cost of longer processing time. It is in this combination that Ddelta can be an excellent substitute to the Xdelta and Zdelta approaches.

7.3   A Deduplication-Aware Elimination Scheme

For delta compression, a key research issue is how to accurately detect a fairly similar candidate for delta compression with low overheads. This section describes DARE, a low-overhead Deduplication-Aware Resemblance detection and Elimination scheme for backup and archiving storage system.

7.3.1   Similarity in Storage System

In a storage system, such as backup system, the modified chunks may be very similar to their previous versions, while unmodified chunks will remain duplicate and are easily identified by the deduplication process. For those non-duplicate chunks that are location-adjacent to known duplicate data chunks in a deduplication system, it is intuitive and quite possible that only a few bytes of them are modified from the last backup, making them potentially excellent delta compression candidates.

Figure 7.6 illustrates a case of duplicate data chunks and their immediate non-duplicate neighbors. As mentioned above, our intuition is that the latter are highly likely to be similar and thus good delta compression candidates. Specifically,

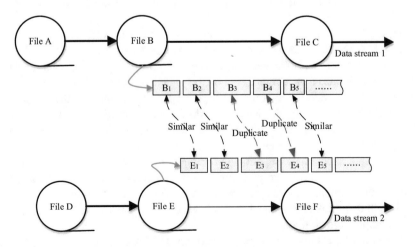

Fig. 7.6 A conceptual illustration of the duplicate adjacency

since chunks B3 and B4 are duplicates of chunks E3 and E4 in Fig. 7.6 respectively, their immediate neighbors, the chunk-pairs B1 and E1, B2, and E2, and B5 and E5, are then considered good delta compression candidates, which is consistent with the aforementioned backup-stream locality [7, 16, 35, 68, 117].

If we can make full use of the existing knowledge about duplicate data chunks in a deduplication system, it is possible for us to detect similar chunks without the overheads of computing and storing features and super-features and then accessing their on-disk index.

7.3.2 Architecture Overview

DARE is designed to improve resemblance detection for additional data reduction in deduplication-based backup/archiving storage systems. As shown in Fig. 7.7, the DARE architecture consists of three functional modules, namely, the Deduplication module, the DupAdj Detection module, and the improved Super-Feature module. In addition, there are five key data structures in DARE, namely, Dedupe Hash Table, SFeature Hash Table, Locality Cache, Container, Segment, and Chunk, which are defined below:

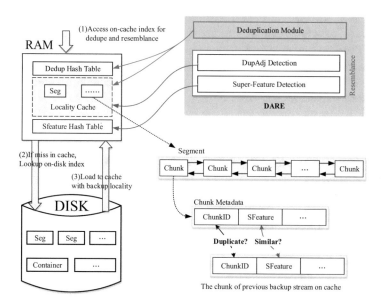

Fig. 7.7 Architecture and key data structures of the DARE system that combines duplicate detection and resemblance detection for data reduction

- A **chunk** is the atomic unit for data reduction. The non-duplicate chunks, identified by their SHA-1 fingerprints, will be prepared for resemblance detection in DARE.
- A **container** is the fixed-size storage unit that stores sequential and NOT reduced data, such as non-duplicate and non-similar or delta chunks, for better storage performance by using large I/Os [16, 35].
- A **segment** consists of the metadata of a number of sequential chunks (e.g., 1 MB size), such as the chunk fingerprints and size, which serves as the atomic unit in preserving the backup-stream logical locality [35] for data reduction. Here DARE uses a data structure of doubly linked list to record the chunk adjacency information for the DupAdj detection. Note that the SFeature in the segment may be unnecessary if the DupAdj module has already confirmed this chunk as being similar for delta compression.
- **Dedupe Hash Table** serves to index fingerprints for duplicate detection for the deduplication module.
- **SFeature Hash Table** serves to index the super-features after the DupAdj resemblance detection. It manages the super-features of non-duplicate and non-similar chunks.
- **Locality Cache** contains the recently accessed data segments and thus preserves the backup-stream locality in memory, to reduce accesses to the on-disk index from either duplicate detection or resemblance detection.

Here we describe a general workflow of DARE. For the input data stream, DARE will first detect duplicate chunks by the Deduplication module. Any of the many existing deduplication approaches [35] can be implemented here and the preservation of the backup-stream logical locality in the segments is required for further resemblance detection. For each non-duplicate chunk, DARE will first use its DupAdj Detection module to quickly determine whether it is a delta compression candidate. If it is not a candidate, DARE will then compute its features and super-features, using its improved Super-Feature Detection module, to further detect resemblance for data reduction.

7.3.3 Duplicate-Adjacency-Based Resemblance Detection Approach

As a salient feature of DARE, the DupAdj approach detects resemblance by exploiting existing duplicate-adjacency information of a deduplication system. The main idea behind this approach is to consider chunk pairs closely adjacent to any confirmed duplicate-chunk pair between two data streams as resembling pairs and thus candidates for delta compression.

According to the description of the DARE data structures in Fig. 7.7, DARE records the backup-stream logical locality of chunk sequence by a doubly linked list. When the DupAdj Detection module of DARE processes an input segment, it will traverse all the chunks by the aforementioned doubly linked list to find the already duplicate-detected chunks. If chunk Am of the input segment A was detected to be a duplicate of chunk Bn of segment B, DARE will traverse the doubly linked list of Bn in both directions (e.g., Am+1 and Bn+1, Am−1 and Bn−1) in search of potentially similar chunk pairs between segments A and B, until a dissimilar chunk or an already detected duplicate or similar chunk is found. Note that the detected chunks here are considered dissimilar (i.e., NOT similar) to others if their similarity degree (i.e., $\frac{dela\ compressed\ size}{chunk\ size}$) is smaller than a predefined threshold, such as 0.25, a false positive for resemblance detection. Actually, the similarity degree of the DupAdj-detected chunks tends to be very high.

In general, the overheads for the DupAdj-based approach are twofold:

- **Memory overhead**: Each chunk will be associated with two pointers (about 8 or 16 Bytes) for building the doubly linked list when DARE loads the segment into the locality cache. But when the segment is evicted from the cache, the doubly linked list will be immediately freed. Therefore, this RAM memory overhead is arguably negligible given the total capacity of the locality cache.
- **Computation overhead**: Confirming the similarity degree of the DupAdj-detected chunks may introduce additional but omitted computation overhead. First, the delta encoding results for the confirmed resembling (i.e., similar) chunks will be directly used as the final delta chunk for storage. Second, the actual extra computation overhead occurs when the DupAdj-detected chunks are NOT similar, which is a very rare event.

7.3.4 Improved Super-Feature Approach

Traditional super-feature approaches generate features by Rabin fingerprints and group these features into super-features to detect resemblance for data reduction. For example, $Feature_i$ of a chunk (length $= N$), is uniquely generated with a randomly predefined value pair mi and ai and N Rabin fingerprints (as used in Content-Defined Chunking [18]) as follows:

$$Feature_i = Max_{j=1}^{N}\left\{\left(m_i * Rabin_j + a_i\right)mod2^{32}\right\} \qquad (7.1)$$

A super-feature of this chunk, $SFeature_x$, can then be calculated by several such features as follows:

$$SFeature_x = Rabin\left(Feature_{x*k}, \ldots, Feature_{x*k+k-1}\right) \qquad (7.2)$$

The less similar two data chunks are to each other, the smaller the probability there will be of them having the same feature. Thus, the probability of two data chunks S1 and S2 being detected as resembling each other by N features can be computed as follows.

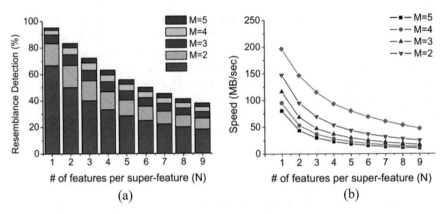

Fig. 7.8 The predicted data reduction efficiency and computation throughput of the super-feature approach as a function of the number of features per super-feature (N, *x*-axis) and the number of super-features (M, segments on each bar in (**a**) or lines in (**b**))

$$Pr\left[\bigcap_{i=1}^{N} max_{i}(H(S_1)) = max_{i}(H(S_2))\right] = \left\{\frac{|\ s_1 \cap s_2\ |}{|\ s_1 \cup s_2\ |}\right\}^{N} = \gamma^{N} \qquad (7.3)$$

This probability is clearly decreasing as a function of the number of features used in a super-feature, as indicated by the above probability expression. Nevertheless, all recent studies on delta compression suggest to increase the number of super-features [7, 32]. If any one of the super-features of two data chunks matches, the two chunks are considered similar to each other. Thus, the probability of resemblance detection, expressed as $1-(1 - \gamma N)M$, can be increased by the number of super-features, M.

For simplicity, assume that the similarity degree γ follows a uniform distribution in the range [0, 1] (note that the actual distribution may be much more complicated in real workloads), the expected value of resemblance detection can be expressed as a function of the number of features per super-feature and the number of super-features under the aforementioned assumption as:

$$\int_{0}^{1} x\left(1 - (1 - x^N)^M\right)dx = \sum_{i=1}^{M} C_M^i (-1)^{i+1} \frac{1}{N * i + 2} \qquad (7.4)$$

This expression of resemblance detection suggests that the larger the number of features used in obtaining a super-feature, N, is, the less capable the super-feature is of resemblance detection. On the other hand, the larger the number of super-features, M, is, the more resemblance can be detected and the more redundancy will be eliminated. Figure 7.8(a) shows the trend of resemblance detection as a function of N and M. Note that the computation overhead of the super-feature-based resemblance approach is proportional to the total number of features N*M, as illustrated in Fig. 7.8(b).

In general, using fewer features per super-feature not only reduces the computation overhead but also detects more resemblance. Thus, DARE employs an improved super-feature approach with fewer features per super-feature and keeps the number of super-features stable to effectively complement the DupAdj resemblance detection.

7.3.5 Delta Compression

To reduce data redundancy among similar chunks, Xdelta [122], an optimized delta compression algorithm, is adopted in DARE after a delta compression candidate is detected by DARE's resemblance detection. In DARE, delta compression will not be applied to a chunk that has already been delta compressed to avoid recursive backward referencing. And DARE records the similarity degree as the ratio of the compressed size to the original size after delta compression (note that "compressed size" here refers to the size of redundant data reduced by delta compression). For example, if delta compression removes 4/5 of data volume in the input chunks detected by DARE, then the similarity degree of the input chunks is 80%, meaning that the volume of the input chunks can be reduced to 1/5 of its original volume by the resemblance detection and delta compression techniques.

Since delta compression needs to frequently read the base-chunks to delta compress the candidate chunks identified by resemblance detection, these frequent disk reads will inevitably slow down the process of data reduction. In order to minimize disk reads, an LRU-based and backup-stream locality-preserved cache of base-chunks is implemented in DARE to load the entire container containing the missing base-chunk to the memory. While our exploitation of the backup-stream locality to prefetch base-chunks can reduce disk reads, some random accesses to on-disk base-chunks are still unavoidable as discussed in [32].

7.3.6 The Detail of DARE Scheme

To put things in perspective, Fig. 7.9 shows a detailed case of the processes of DARE system. "D," "S," and "N" in the figure refer to a duplicate chunk, a similar chunk, and a chunk that is neither duplicate nor similar, respectively.

For an incoming backup stream, DARE goes through the following four key steps:

1. **Duplicate Detection.** The data stream is first chunked, fingerprinted, duplicate-detected, and then grouped into segments of sequential chunks to preserve the backup-stream logical locality [35]. Note that the locality information will be exploited by the following DupAdj resemblance detection.

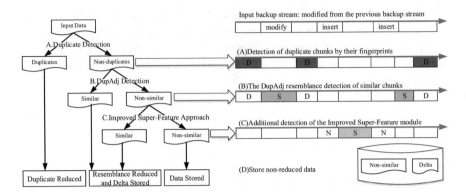

Fig. 7.9 The data reduction workflow of DARE, showing an example of resemblance detection for delta compression first by the DupAdj approach and then by the super-feature approach

2. **Resemblance Detection.** The DupAdj resemblance detection module in DARE first detects duplicate-adjacent chunks in the segments formed in step (1). After that, DARE's improved super-feature module further detects similar chunks in the remaining non-duplicate and non-similar chunks that may have been missed by the DupAdj detection module when the duplicate-adjacency information is lacking or weak.

3. **Delta Compression.** For each of the resembling chunks detected in step (2), DARE reads its base-chunk, then delta encodes their differences. In order to reduce disk reads, an LRU and locality-preserved cache is implemented here to prefetch the base-chunks in the form of data segments.

4. **Storage Management.** The data NOT reduced, i.e., non-similar and delta chunks, will be stored as containers on the disk. The file mapping relationships among the duplicate chunks, resembling chunks, and non-similar chunks will also be recorded as the file recipes [71,83] to facilitate future data restore operations in DARE.

For the restore operation, DARE will first read the referenced file recipes and then read the duplicate and non-similar chunks one by one from the referenced segments on disk according to mapping relationships in the file recipes. For the resembling chunks, DARE needs to read both delta data and base-chunks and then delta decode them to the original ones.

7.3.7 Performance Evaluation

Platform of the DARE Prototype. We have implemented a prototype of DARE and tested it on the Ubuntu 12.04 operating system running on a quad-core Xeon E5606 processor at 2.13 GHz, with a 16GB RAM, a 14TB RAID6 disk array that consists of sixteen 1TB disks, and a 120GB SSD of KINGSTON SVP200S37A120G.

Table 7.4 Workload characteristics of six open-source project datasets, two synthetic backup datasets, and one database dataset used in the performance evaluation

Datasets	Versions	Size
Emacs-21.4~Emacs-23.4	8	1.15 GB
GDB-6.7~GDB-7.4.1	10	1.37 GB
Glibc-2.1.1~Glibc-2.15	35	3.18 GB
SciLab-5.0.1~SciLab-5.3.2	10	4.94 GB
GCC-4.3.4~GCC-4.7.0	20	8.91 GB
Linux-3.0.0~Linux-3.0.39	40	16.8 GB
Freq (20% inc. of newer version)	20	857 GB
Less (10% inc. of newer version)	30	1372 GB
RDB (backups of Redis database)	100	540 GB

Configurations for Data Reduction. DARE employs the widely used Rabin algorithm [61,128] and SHA-1 hash function [16] respectively for chunking and fingerprinting for data deduplication, and an average chunk size of 8KB. For the resemblance detection, DARE adopts a CDC sliding window size of 48 bytes to generate features and Xdelta [122] to compress the detected similar chunks.

Datasets. Six well-known open-source projects [77,133,136] representing typical workloads of deduplication and resemblance detection are used in the evaluation of DARE as shown in Table 7.4. These datasets consist of large tarred files representing sets of source code files or objects concatenated together by backup software [6].

Evaluation of DARE. We evaluate DARE's resemblance detection and elimination schemes, i.e., the DupAdj resemblance detection and the improved super-feature approach with three SFs and two features per SF., We compare DARE with the super-feature-only approaches based on three SFs with two-features per SF (SF-2F) and three SFs with four-features per SF (SF-4F). Note that DARE's resemblance detection here is thus DupAdj supplemented by SF-2F (i.e., DARE = DupAdj + SF-2F), where SF-2F is applied only to chunks that DupAdj has failed to detect as being similar.

Table 7.5 shows the additional data reduction on top of the conventional deduplication (Dedupe) achieved by the four resemblance detection schemes, DupAdj, DARE, SF-2F, and SF-4F, on the six real-world datasets (both tarred and untarred). "+" in the figure denotes additional data reduction beyond deduplication. Generally, the DupAdj approach achieves a resemblance-detection efficiency similar to the SF-4F approach and DARE detects about 2–6% and 3–10% more redundancy than the SF-2F and SF-4F approaches respectively. Thus DARE detects the most resemblance by combining the DupAdj and SF-2F resemblance detection approaches.

Figure 7.10(a) shows that the average similarity degree of the resembling chunks detected by the DARE, SF-2F, and SF-4F approaches is about 0.890, 0.922, and 0.965 respectively. Therefore, the results demonstrate that DupAdj and DARE are very effective and efficient in detecting resemblance among the post-deduplication chunks with a very low false-positive rate.

Figure 7.10b, c show the computation and indexing overheads incurred by the three resemblance detection schemes. Obviously, SF-4F, which computes more

Table 7.5 Comparisons among Deduplication, DupAdj, DARE, SF-2F, and SF-4F approaches in the data reduction measure under six real-world datasets, both tarred and untarred versions for a total of 12 datasets

Datasets		Versions	Dedupe	DupAdj	DARE	SF-2F	SF-4F
	Emacs	8	37.1%	+32.1%	+41.0%	+33.7%	+28.2%
	GDB	10	48.7%	+33.5%	+40.8%	+36.4%	+33.4%
	Glibc	35	52.2%	+29.2%	+36.9%	+35.3%	+30.4%
Tarred	SciLab	10	56.9%	+19.5%	+25.2%	+22.6%	+18.8%
	GCC	20	39.1%	+38.2%	+46.7%	+45.2%	+40.6%
	Linux	40	40.9%	+53.4%	+54.1%	+54.4%	+53.5%
	Emacs	8	43.5%	+29.6%	+37.7%	+31.7%	+28.0%
	GDB	10	70.6%	+10.9%	+18.2%	+16.5%	+14.2%
	Glibc	35	87.9%	+2.9%	+7.3%	+6.6%	+5.7%
UnTarred	SciLab	10	77.5%	+5.2%	+10.4%	+9.6%	+8.1%
	GCC	20	83.5%	+7.2%	+9.9%	+9.1%	+7.3%
	Linux	40	96.7%	+0.7%	+1.0%	+0.9%	+0.6%

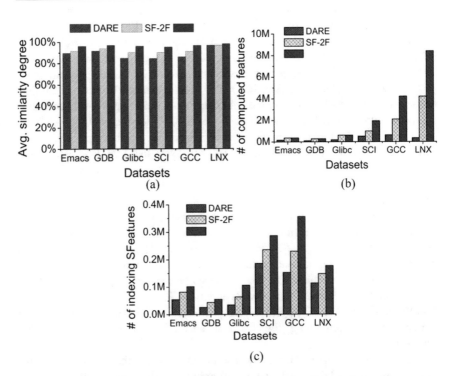

Fig. 7.10 Comparison between DARE, SF-2F, and SF-4F in terms of the similarity degree and the computation and indexing overheads. (**a**) Average similarity degree. (**b**) Computation overhead. (**c**) Indexing overhead

Table 7.6 Comparisons among the duplicate-detection and resemblance-detection approaches in terms of data reduction efficiency on three larger backup datasets Freq, Less, and RDB

Datasets	Freq		Less		RDB	
Dedupe	84.6%	6.5×	91.2%	11.4×	95.7%	23.2
SiLo/D	84.4%	6.4×	91.2%	11.4×	95.7%	23.2
Cac. DARE	+9.28%	*2.5×	+5.01%	*2.3×	+2.97%	*3.2×
Cac. SF-2F	+7.01%	*1.8×	+4.31%	*2.0×	+2.95%	*3.1×
Cac. SF-4F	+5.32%	*1.5×	+3.27%	*1.6×	+2.82%	*2.9×
Ful. SF-2F	+9.61%	*2.6×	+5.04%	*2.3×	+2.97%	*3.2×

features but detects less resemblance, consumes the most amounts of computation and indexing resources for resemblance detection. DARE uses the same super-feature parameters as SF-2F but incurs only half of the computation and indexing overheads of the SF-2F approach because of DupAdj's very effective prescreening of similar chunks. In fact, DARE can further reduce the number of super-features while achieving a comparable resemblance detection efficiency to the SF-2F approach.

Next we evaluate the scalability of DARE. In order to better evaluate the scalability of DARE on three larger datasets, we have implemented the schemes of Stream-Informed Delta Compression (SIDC) [7] in SiLo [117], a memory-efficient deduplication system that exploits the backup-stream similarity and locality. SIDC only detects resemblance in the backup-stream locality-preserved cache that can reduce the indexing overhead of SFs and scales well in large-scale deduplication system. Thus we employ their method to test the scalability of different resemblance detection schemes and implement SiLo with a 20MB locality cache (similar to SIDC [7]) and a segment size of 1MB. The stream-informed approaches are denoted by the "Cached/Cac." prefix in Table 7.6 and Fig. 7.11.

Table 7.6 shows the data reduction results of different deduplication and resemblance detection schemes. "+" ("*") sign in front of a reduction percentage (factor) in the table indicates the "additional" post-deduplication data reduction. We can see that resemblance detection further reduces the storage space, after the deduplication process, by a factor of about 2.5, meaning that we can save about 60% of the post-deduplication storage space by resemblance-detecting post-deduplication chunks. Figure 7.11 shows that DARE achieves the highest throughput among all the resemblance detection enhanced data reduction approaches compared running on both RAID-structured HDDs and the SSD. DARE achieve lower throughput than the SiLo/D approach, which is because SiLo/D only does deduplication (i.e., Rabin-based chunking and SHA1-based fingerprinting) while DARE involves more computation tasks and I/Os by delta compression (i.e., resemblance detection, reading base chunks, and delta encoding). In general, DARE achieves a superior performance of both throughput and data reduction efficiency among all the resemblance detection enhanced data reduction approaches.

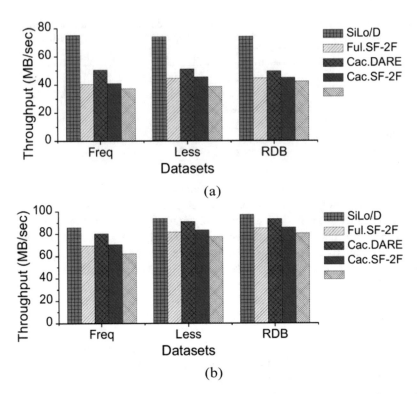

Fig. 7.11 Throughput of four resemblance detection enhanced data reduction approaches on the three datasets. (**a**)Throughput on RAID. (**b**)Throughput on SSD

7.4   Concluding Remarks

Data deduplication technology often fails to identify redundancy among non-duplicate but very similar chunks, but delta compression is able to make up for that. In this chapter we present two post-deduplication delta compression schemes, Ddelta and DARE. Ddelta is a deduplication-inspired fast delta compression scheme that effectively leverages the principles of deduplication to improve delta encoding and decoding speeds without sacrificing compression ratio. DARE uses a novel approach, DupAdj, which exploits the duplicate-adjacency information for efficient resemblance detection in existing deduplication systems, and employs an improved super-feature approach to further detecting resemblance when the duplicate-adjacency in-formation is lacking or limited.

Chapter 8
The Framework of Data Deduplication

Abstract Some open-source deduplication systems or frameworks have been released, but they did not achieve a complete parameter space, so the researchers cannot conduct experiments on different design comparison. In order to test the parameter space discussed above and find a reasonable parameter configuration, we present the design and implement details of our deduplication prototype, DeFrame (or destor). We implement a DeFrame prototype in 10000+ C codes, which is installed with Linux 64-bit operating system and depend on pthread, OpenSSL, and Glib libraries. The source code of our REED prototype is available for download at https://github.com/fomy/destor. The rest of this chapter is organized as follows. Section 8.1 describes the parameter space of in-line data deduplication. Section 8.2 presents the architecture and submodules of DeFrame. Section 8.3 proposes lots of experiments on real-world datasets to discuss the relationship and the configuration of parameter space. Section 8.4 gives some design recommendation based on different goals.

8.1 In-Line Data Deduplication Space

In the following, we (1) propose the fingerprint index subspace (the key component in data deduplication systems) to characterize existing solutions and find potentially better solutions, and (2) discuss the interplays among fingerprint index, rewriting, and restore algorithms. Table 8.1 lists the major parameters.

The fingerprint index is a well-recognized performance bottleneck in large-scale deduplication systems [16]. The simplest fingerprint index is only a key-value store [17]. The key is a fingerprint and the value points to the chunk. A duplicate chunk is identified via checking the existence of its fingerprint in the key-value store.

An HDD-based key-value store suffers from HDD's poor random-access performance, since the fingerprint is completely random in nature. For example, the throughput of Content-Defined Chunking (CDC) is about 400 MB/s under commercial CPUs [134], and hence CDC produces 102, 400 chunks per second. Each chunk incurs a lookup request to the key-value store, i.e., 102, 400 lookup requests per second. The required throughput is significantly higher than that of HDDs, i.e.,

© Springer Nature Singapore Pte Ltd. 2022
D. Feng, *Data Deduplication for High Performance Storage System*,
https://doi.org/10.1007/978-981-19-0112-6_8

Table 8.1 The major parameters we discuss

Parameter list	Description
Sampling	Selecting representative fingerprints
Segmenting	Splitting the unit of logical locality
Segment selection	Selecting segments to be prefetched
Segment prefetching	Exploiting segment-level locality
Key-value mapping	Multiple logical positions per fingerprint
Rewriting algorithm	Reducing fragmentation
Restore algorithm	Designing restore cache

100 IOPS [73]. SSDs support much higher throughput, nearly 75,000 as venders reported [135]. However, SSDs are much more expensive than HDDs and suffer from a performance degradation over time due to reduced over-provisioning space [136].

Due to the incremental nature of backup workloads, the fingerprints of consecutive backups appear in similar sequences [16], which is known as *locality*. In order to reduce the overhead of the key-value store, modern fingerprint indexes leverage locality to prefetch fingerprints, and maintain a *fingerprint cache* to hold the prefetched fingerprints in memory. The fingerprint index hence consists of two sub-modules: a key-value store and a fingerprint prefetching/caching module. The value instead points to the prefetching unit. According to the use of the key-value store, we classify the fingerprint index into *exact* and *near-exact deduplication*.

- Exact deduplication: all duplicate chunks are eliminated for highest deduplication ratio (the data size before deduplication divided by the data size after deduplication).
- Near-exact deduplication: a small number of duplicate chunks are allowed for higher backup performance and lower memory footprint.

  According to the fingerprint prefetching policy, we classify the fingerprint index into exploiting *logical* and *physical locality*.
- Logical locality: The chunk (fingerprint) sequence of a backup stream, namely the chunk sequence before deduplication. It is preserved in recipes.
- Physical locality: The physical sequence of chunks (fingerprints), namely the chunk sequence after deduplication. It is preserved in containers.

Figure 8.1 shows the categories of existing fingerprint indexes. The cross-product of the deduplication and locality variations include EDPL, EDLL, NDPL, and NDLL. The typical examples include DDFS [16], Sparse Indexing [68], Extreme Binning [71], ChunkStash [73], Sampled Index [23], SiLo [116], PRUNE [69], and BLC [137].

Fig. 8.1 Categories of existing fingerprint indexes

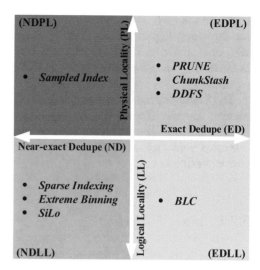

8.2 Framework Architecture

As shown in Fig. 8.2, DeFrame consists of three sub-modules: fingerprint index, container store, and recipe store[35]. In the *container store*, each container is identified by a globally unique ID. The container is the prefetching unit for exploiting physical locality. Each container includes a metadata section that summarizes the fingerprints of all chunks in the container. We can fetch an entire container or only its metadata section via a container ID.

The *recipe store* manages recipes of all finished backups. In recipes, the associated container IDs are stored along with fingerprints so as to restore a backup without the need to consult the fingerprint index. We add some indicators of segment

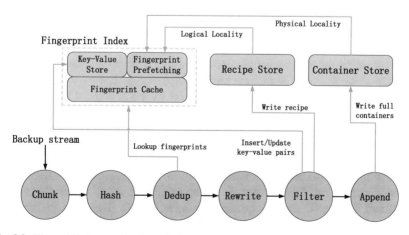

Fig. 8.2 The architecture and backup pipeline

boundaries in each recipe to facilitate reading a segment that is the prefetching unit for exploiting logical locality. Each segment is identified by a globally unique ID. For example, an ID can consist of a 2-byte pointer to a recipe, a 4-byte offset in the recipe, and a 2-byte segment size that indicates how many chunks are in the segment.

The *fingerprint index* consists of a key-value store and a fingerprint prefetching/ caching module. Two kinds of key-value stores are currently supported: an in-DRAM hash table and a MySQL database [138] paired with a Bloom filter. Since we implement a virtual layer upon the key-value store, it is easy to add a new key-value store.

8.2.1   Backup Pipeline

As shown in Fig. 8.2, we divide the workflow of data deduplication into six phases: *Chunk, Hash, Dedup, Rewrite, Filter,* and *Append.* (1) The Chunk phase divides the backup stream into chunks. We have implemented Fixed-Sized Chunking and Content-Defined Chunking (CDC). (2) The Hash phase calculates a SHA-1 digest for each chunk as the fingerprint. (3) The Dedup phase aggregates chunks into segments, and identifies duplicate chunks via consulting the fingerprint index. A duplicate chunk is marked and obtains the container ID of its stored copy. The created segments are the prefetching units of logical locality, and the batch process units for physical locality. We have implemented the Base, Top-k, and *Mix* procedures (first Top-k then Base). (4) The Rewrite phase identifies fragmented duplicate chunks, and rewrites them to improve restore performance. It is a trade-off between deduplication ratio and restore performance. We have implemented four rewriting algorithms, including CFL-SD [58], CBR [82], Capping [83], and HAR [139]. Each fragmented chunk is marked. (5) The Filter phase handles chunks according to their marks. Unique and fragmented chunks are added to the container buffer. Once the container buffer is full, it is pushed to the next phase. The recipe store and key-value store are updated. (6) The Append phase writes full containers to the container store.

We pipeline the phases via pthreads to leverage multi-core architecture. The dedup, rewrite, and filter phases are separated for modularity: we can implement a new rewriting algorithm without the need to modify the fingerprint index, and vice versa.

Segmenting and Sampling. The segmenting method is called in the dedup phase, and the sampling method is called for each segment either in the dedup phase for the similarity detection or in the filter phase for the Base procedure. All segmenting and sampling methods have been implemented. The segmenting methods consist of File-Defined Segmenting (FDS), Fixed-Sized Segmenting (FSS) and Content-Defined Segmenting method (CDS). The sampling methods consist of three basic sampling methods: *uniform, random,* and *minimum.* Content-defined segmenting is implemented via checking the last n bits of a fingerprint. If all the bits are zero, the fingerprint (chunk) is considered to be the beginning of a new segment, thus generating an average segment size of 2^n chunks. To select the first fingerprint of a content-defined segment as a feature, the random sampling also checks the last $\log_2 R$ ($<n$) bits.

8.2.2   Restore Pipeline

The restore pipeline in DeFrame consists of three phases: *Reading Recipe*, *Reading Chunks*, and *Writing Chunks*.

1. Reading Recipe. The required backup recipe is opened for restore. The finger-prints are read and issued one by one to the next step.
2. Reading Chunks. Each fingerprint incurs a chunk read request. The container is read from the container store to satisfy the request. A chunk cache is maintained to hold popular chunks in memory. We have implemented three kinds of restore algorithms, namely the basic LRU cache, the optimal cache [139], and the rolling forward assembly area [83]. Given a chunk placement determined by the rewrit-ing algorithm, a good restore algorithm boosts the restore procedure with a limited memory footprint. The required chunks are issued one by one to the next phase.
3. Writing Chunks. Using the received chunks, files are reconstructed in the local file system.

8.2.3   Garbage Collection

After users delete expired backups, chunks become invalid (not referenced by any backup) and must be reclaimed. There are a number of possible techniques for *garbage collection* (GC), such as reference counting [140] and mark-and-sweep [23]. Extending the DeFrame taxonomy to allow comparison of GC techniques is beyond the scope of this work; currently, DeFrame employs the History-Aware Rewriting (HAR) algorithm and Container-Marker Algorithm (CMA) proposed in [139]. HAR rewrites fragmented valid chunks to new containers during backups, and CMA reclaims old containers that are no longer referenced.

8.3   Performance Evaluation

8.3.1   Experimental Setup

We use three real-world datasets as shown in Table 8.2. Kernel is downloaded from the web [87]. It consists of 258 versions of unpacked Linux kernel source code.

Table 8.2 The characteristics
of datasets

Dataset name	Kernel	VMDK	RDB
Total size	104 GB	1.89 TB	1.12 TB
# of versions	258	127	212
Deduplication ratio	45.28	27.36	39.1
Avg. chunk size	5.29 KB	5.25 KB	4.5 KB
Self-reference	1%	15–20%	0
Fragmentation	Severe	Moderate	Severe

VMDK is from a virtual machine with Ubuntu 12.04. We compiled the source code, patched the system, and ran an HTTP server on the virtual machine. VMDK has many self-references; it also has less fragmentation from its fewer versions and random updates. RDB consists of Redis database [131] snapshots. The database has 5 million records, 5 GB in space, and an average 1% update ratio. We disable the default rdb compression option.

All datasets are divided into variable-sized chunks via CDC. We use the content-defined segmenting with an average segment size of 1024 chunks by default. The container size is 4 MB, which is close to the average size of segments. The default fingerprint cache has 1024 slots to hold prefetching units, being either containers or segments. Hence, the cache can hold 1 million fingerprints, which is relatively large for our datasets.

8.3.2   Metrics and Our Goal

Our evaluations are in terms of quantitative metrics listed as follows. (1) *Deduplication ratio*: the original backup data size divided by the size of stored data. It indicates how efficiently data deduplication eliminates duplicates, being an important factor in the storage cost. (2) *Memory footprint*: the runtime DRAM consumption. A low memory footprint is always preferred due to DRAM's high unit price and energy consumption. (3) *Storage cost*: the cost for storing chunks and the fingerprint index, including memory footprint. We ignore the cost for storing recipes, since it is constant. (4) *Lookup requests per GB*: the number of required lookup requests to the key-value store to deduplicate 1 GB of data, most of which are random reads. (5) *Update requests per GB*: the number of required update requests to the key-value store to deduplicate 1 GB of data. A higher lookup/update overhead degrades the backup performance. Lookup requests to unique fingerprints are eliminated since most of them are expected to be answered by the in-memory Bloom filter. (6) *Restore speed*: 1 divided by mean containers read per MB of restored data [83]. It is used to evaluate restore performance, where a higher value is better. Since the container size is 4 MB, 4 units of restore speed translate to the maximum storage bandwidth.

It is practically impossible to find a solution that performs the best in all metrics. Our goal is to find some reasonable solutions with the following properties:

(1) sustained, high backup performance as the top priority; (2) reasonable trade-offs in the remaining metrics.

8.3.3 Exact Deduplication

Previous studies [16, 73] of EDPL fail to have an insight of the impacts of the fragmentation on the backup performance, since their datasets are short-term. Figure 8.3 shows the ever-increasing lookup overhead. $R = 256$ indicates a sampling ratio of 256:1. Results come from RDB. In Fig. 8.3(a), FSS is Fixed-Sized Segmenting and CDS is Content-Defined Segmenting. Points in a line are of different sampling ratios, which are 256, 128, 64, 32, and 16 from left to right. In Fig. 8.3(b), EDLL is of CDS and a 256:1 random sampling ratio. We observe 6.5–12.0× and 5.1–114.4× increases in Kernel and RDB respectively under different fingerprint cache sizes. A larger cache cannot address the fragmentation problem; a 4096-slot cache performs as poor as the default 1024-slot cache. A 128-slot cache results in a 114.4× increase in RDB, which indicates that an insufficient cache can result in unexpectedly poor performance. This causes complications in practice due to the difficulty in predicting how much memory is required to avoid unexpected performance degradations. Furthermore, even with a large cache, the lookup overhead still increases over time.

Before comparing EDLL to EDPL, we need to determine the best segmenting and sampling methods for EDLL. Figure 8.4 shows the lookup/update overheads of EDLL under different segmenting and sampling methods in VMDK. Similar results are observed in Kernel and RDB. Increasing the sampling ratio shows an efficient trade-off: a significantly lower update overhead at a negligible cost of a higher lookup overhead. The fixed-sized segmenting paired with the random sampling performs worst. This is because it cannot sample the first fingerprint in a segment, which is important for the Base procedure. The other three combinations are more efficient since they sample the first fingerprint (the random sampling performs well in the content-defined segmenting due to our optimization). The content-defined

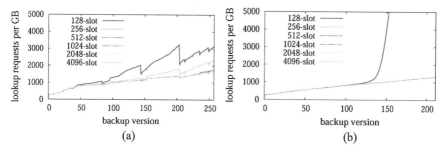

Fig. 8.3 The ever-increasing lookup overhead of EDPL in Kernel and RDB under various fingerprint cache sizes. (**a**) Kernel. (**b**) RDB

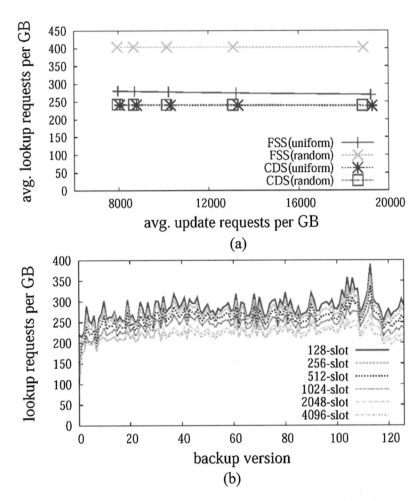

Fig. 8.4 Impacts of varying segmenting, sampling, and cache size on EDLL in VMDK. (**a**) Segmenting and sampling. (**b**) Cache size

segmenting is better than the fixed-sized segmenting due to its shift-resistance. Figure 8.4(b) shows the lookup overheads in VMDK under different cache sizes. We do not observe an ever-increasing trend of lookup overhead in EDLL. A 128-slot cache results in additional I/O (17% more than the default) due to the space-oriented similar segments in VMDK. Kernel and RDB (not shown in the figure) do not cause this problem because they have no self-reference.

Figure 8.5 compares EDPL and EDLL in terms of lookup and update overheads. EDLL uses the content-defined segmenting and random sampling. Results in Kernel and VMDK are not shown, because they have results similar to RDB. While EDPL suffers from the ever-increasing lookup overhead, EDLL has a much lower and sustained lookup overhead (3.6× lower than EDPL on average). With a 256:1

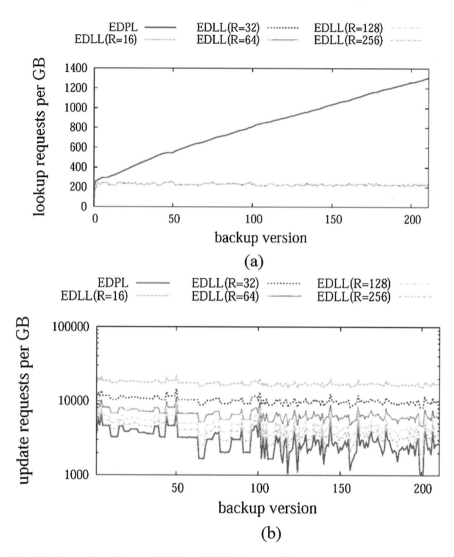

Fig. 8.5 Comparisons between EDPL and EDLL in terms of lookup and update overheads. (**a**) Lookup overhead. (**b**) Update overhead

sampling ratio, EDLL has $1.29\times$ higher update overhead since it updates sampled duplicate fingerprints with their new segment IDs. Note that lookup requests are completely random, and update requests can be optimized to sequential writes via a log-structured key-value store, which is a popular design [141–143]. Overall, if the highest deduplication ratio is required, EDLL is a better choice due to its sustained high backup performance.

Finding (1): While the fragmentation results in an ever-increasing lookup over-head in EDPL, EDLL achieves sustained performance. The sampling optimization performs an efficient trade-off in EDLL.

8.3.4 Near-Exact Deduplication Exploiting Physical Locality

NDPL is simple and easy to implement. Figure 8.6(a) shows how to choose an appropriate sampling method for NDPL. Points in each line are of different sampling ratios, which are 256, 128, 64, 32, 16, and 1 from left to right. The Y-axis shows the

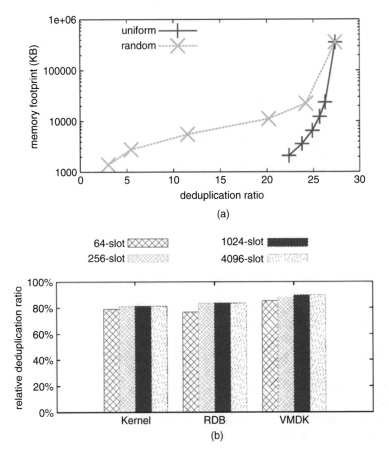

Fig. 8.6 Impacts of varying sampling methods and cache sizes on NDPL. (**a**) Sampling. (**b**) Cache size

relative deduplication ratio to exact deduplication. We only show the results from VMDK, which are similar to the results from Kernel and RDB.

The uniform sampling achieves significantly higher deduplication ratio than the random sampling. For the random sampling, the missed duplicate fingerprints are definitely not sampled, making new containers have less features and hence smaller probability of being prefetched. The sampling ratio is a trade-off between memory footprint and deduplication ratio: a higher sampling ratio indicates a lower memory footprint at a cost of a decreased deduplication ratio. Figure 8.6(b) shows that NDPL is surprisingly resistant to small cache sizes: a 64-slot cache results in only an 8% decrease of the deduplication ratio than the default in RDB. NDPL uses a 128:1 uniform sampling ratio. The Y-axis shows the relative deduplication ratio to exact deduplication. Also observed (not shown in the figure) are 24–93% additional I/O, which come from prefetching fingerprints. Compared to EDPL, NDPL has better backup performance because of its in-memory key-value store at a cost of decreasing deduplication ratio.

> *Finding (2): In NDPL, the uniform sampling is better than the random sampling. The fingerprint cache has minimal impacts on deduplication ratio.*

8.3.5 Near-Exact Deduplication Exploiting Logical Locality

Figure 8.7(a) compares the Base procedure to the simplest similarity detection Top-1, which helps to choose the appropriate sampling method. Points in each line are of different sampling ratios, which are 512, 256, 128, 64, 32, and 16 from left to right. The content-defined segmenting is used due to its advantage shown in EDLL. In the Base procedure, random sampling achieves comparable deduplication ratio using less memory than uniform sampling. NDLL is expected to outperform NDPL in terms of deduplication ratio since NDLL does not suffer from fragmentation. However, we surprisingly observe that, while NDLL does better in Kernel and RDB as expected, NDPL is better in VMDK (shown in Figs. 8.6(b) and 8.7(b)).

NDLL is of the Base procedure and a 128:1 random sampling. The Y-axis shows the relative deduplication ratio to exact deduplication. The reason is that self-reference is common in VMDK. The fingerprint prefetching is misguided by space-oriented similar segments as discussed. Moreover, the fingerprint cache contains many duplicate fingerprints that reduce the effective cache size, therefore a 4096-slot cache improves deduplication ratio by 7.5%. NDPL does not have this problem since its prefetching unit (i.e., container) is after-deduplication. A 64-slot cache results in 23% additional I/Os in VMDK (not shown in the figure), but has no side-effect in Kernel and RDB.

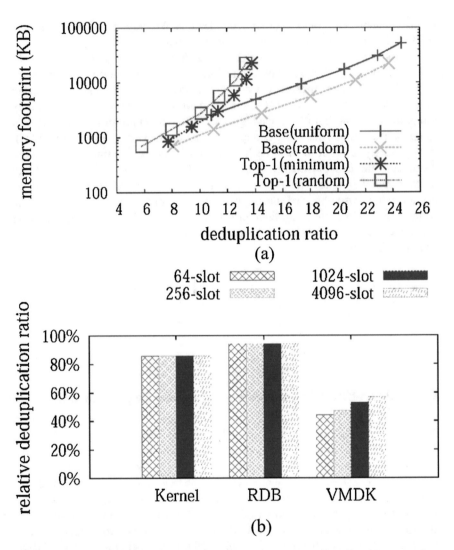

Fig. 8.7 Impacts of varying segment selections and sampling methods (Top-k) and the comparison of different cache sizes on deduplication ratio. (**a**) Segment selection and sampling. (**b**) Cache size

In the Top-1 procedure, only the most similar segment is read. The minimum sampling is slightly better than the random sampling. The Top-1 procedure is worse than the Base procedure. The reason is two-fold as discussed: (1) a segment boundary change results in more time-oriented similar segments; (2) self-reference results in many space-oriented similar segments.

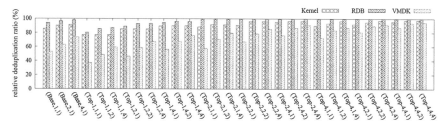

Fig. 8.8 Impacts of the segment selection, segment prefetching, and mapping relationship on deduplication ratio.

Finding (3): The Base procedure underperforms in NDLL if self-reference is common. Reading a single most similar segment is insufficient due to self-reference and segment boundary changes.

We further examine the remaining NDLL subspace: segment selection (s), segment prefetching (p), and mapping relationship (v). Figure 8.8 shows the impacts of varying the three parameters on deduplication ratio (lookup overheads are omitted due to space limits). The deduplication ratios are relative to those of exact deduplication. On the X-axis, we have parameters in the format (s, p, v). The s indicates the segment selection method, being either Base or Top-k. The p indicates the number of prefetched segments plus the selected segment. We apply segment prefetching to all similar segments selected. The v indicates the maximum number of segments that a feature refers to. Random sampling is used, with a sampling ratio of 128. For convenience, we use NDLL (s, p, v) to represent a point in the space.

A larger v results in a higher lookup overhead when $k > 1$, since it provides more similar segment candidates. We observe that increasing v is not cost-effective in Kernel which lacks self-reference, since it increases lookup overhead without an improvement of deduplication ratio. However, in RDB which also lacks of self-reference, NDLL(Top-1,1,2) achieves better deduplication ratio than NDLL (Top-1,1,1) due to the rollbacks in RDB. A larger v is helpful to improve deduplication ratio in VMDK where self-reference is common. For example, NDLL(Top-1,1,2) achieves $1.31\times$ higher deduplication ratio than NDLL (Top-1,1,1) without an in-crease of lookup overhead.

The segment prefetching is efficient for increasing deduplication ratio and decreasing lookup overhead. As the parameter p increases from 1 to 4 in the Base procedure, the deduplication ratios increase by $1.06\times$, $1.04\times$, and $1.39\times$ in Kernel, RDB, and VMDK respectively, while the lookup overheads decrease by $3.81\times$, $3.99\times$, and $3.47\times$. The Base procedure is sufficient to achieve a high deduplication ratio in Kernel and RDB that lack self-reference. Given its simple logical frame, the Base procedure is a reasonable choice if self-reference is rare. However, the Base procedure only achieves a 73.74% deduplication ratio of exact deduplication in VMDK where self-reference is common.

> *Finding (4): If self-reference is rare, the Base procedure is sufficient for a high deduplication ratio.*

In more complicated environments like virtual machine storage, the Top-k procedure is desired. A higher k indicates a higher deduplication ratio at a cost of a higher lookup overhead. As k increases from NDLL(Top-1,1,1) to NDLL (Top-4,1,1), the deduplication ratios increase by 1.17×, 1.24×, and 1.97× in Kernel, RDB, and VMDK respectively, at a cost of 1.15×, 1.01×, and 1.56× more segment reads. Note that Top-4 outperforms Base in terms of deduplication ratio in all datasets. Varying k has fewer impacts in Kernel and RDB, since they have fewer space-oriented similar segments and hence fewer candidates. The segment prefetching is a great complement to the Top-k procedure, since it amortizes the additional lookup overhead caused by increasing k. NDLL(Top-4,4,1) reduces the lookup overheads of NDLL(Top-4,1,1) by 2.79×, 3.97×, and 2.07 × in Kernel, RDB, and VMDK respectively. It also improves deduplication ratio by a factor of 1.2× in VMDK. NDLL(Top-4,4,1) achieves a 95.83%, 99.65%, and 87.20% deduplication ratio of exact deduplication, significantly higher than NDPL.

> *Finding (5): If self-reference is common, the similarity detection is required. The segmenting prefetching is a great complement to Top-k.*

8.3.6   Rewriting Algorithm and Its Interplay

Fragmentation decreases restore performance significantly in aged systems. The rewriting algorithm is proposed to trade deduplication ratio for restore performance. To motivate the rewriting algorithm, Table 8.3 compares near-exact deduplication to a rewriting algorithm, History-Aware Rewriting algorithm (HAR) [139]. Specifically, NDPL-256 indicates NDPL of a 256:1 uniform sampling ratio. HAR uses EDPL as the fingerprint index. The restore cache contains 128 containers for Kernel, and 1024 containers for RDB and VMDK.

We choose HAR due to its accuracy in identifying fragmentation. As the baseline, EDPL has best deduplication ratio and hence worst restore performance. NDPL shows its ability of improving restore performance, however not as well as HAR. Taking RDB as an example, NDPL of a 512:1 uniform sampling ratio trades 33.88% deduplication ratio for only 1.18× improvement in restore speed, while HAR trades 27.69% for 2.8× improvement.

Table 8.3 Comparisons between near-exact deduplication and rewriting in terms of restore speed and deduplication ratio

	Dataset	Deduplication ratio	Restore speed
EDPL	Kernel	45.35	0.50
	RDB	39.10	0.50
	VMDK	27.36	1.39
NDPL-128	Kernel	36.86	0.92
	RDB	32.64	0.82
	VMDK	24.50	2.49
NDPL-256	Kernel	33.53	1.04
	RDB	29.31	0.87
	VMDK	23.15	2.62
NDPL-512	Kernel	31.26	1.19
	RDB	25.86	0.95
	VMDK	21.46	2.74
HAR+EDPL	Kernel	31.26	2.60
	RDB	28.28	2.25
	VMDK	24.90	2.80

We now answer the questions in rewriting: (1) How does the rewriting algorithm improve EDPL in terms of lookup overhead? (2) How does fingerprint index affect the rewriting algorithm? Figure 8.9(a) shows how HAR improves EDPL. We observe that HAR successfully stops the ever-increasing trend of lookup overhead in EDPL. Although EDPL still has a higher lookup overhead than EDLL, it is not a big deal because a predictable and sustained performance is the main concern. Moreover, HAR has no impact on EDLL, since EDLL does not exploit physical locality that HAR improves. The periodic spikes are because of major updates in Linux kernel, such as from 3.1 to 3.2. These result in many new chunks, which reduce logical locality. Figure 8.9(b) shows how fingerprint index affects HAR. The Y-axis shows the relative deduplication ratio to that of exact deduplication without rewriting. EDPL outperforms EDLL in terms of deduplication ratio in all datasets. As explained in rewriting, EDLL could return an obsolete container ID if an old segment is read, and hence a recently rewritten chunk would be rewritten again. Overall, with an efficient rewriting algorithm, EDPL is a better choice than EDLL due to its higher deduplication ratio and sustained performance.

> *Finding (6): The rewriting algorithm helps EDPL to achieve sustained backup performance. With a rewriting algorithm, EDPL is better due to its higher deduplication ratio than other index schemes.*

We further examine three restore algorithms: the LRU cache, the forward assembly area (ASM) [83], and the optimal cache (OPT) [139]. Figure 8.10 shows the efficiencies of these restore algorithms with and without HAR in Kernel and VMDK.

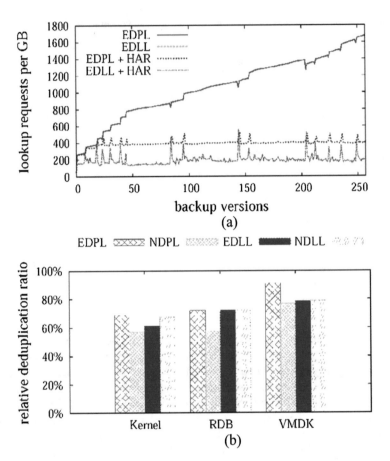

Fig. 8.9 Interplays between fingerprint index and rewriting algorithm (i.e., HAR). (**a**) How does HAR improve EDPL in terms of lookup overhead in Kernel? (**b**) How does fingerprint index affect HAR?

In particular, EDPL is used as the fingerprint index. Because the restore algorithm only matters under limited memory, the DRAM used is smaller than Table 8.3, 32-container-sized in Kernel and 256-container-sized in VMDK. If no rewriting algorithm is used, the restore performance of EDPL decreases over time due to the fragmentation. ASM has better performance than LRU and OPT, since it never holds useless chunks in memory. If HAR is used, EDPL has sustained high restore performance since the fragmentation has been reduced. OPT is best in this case due to its efficient cache replacement.

Finding (7): Without rewriting, the forward assembly area is recommended; but with an efficient rewriting algorithm, the optimal cache is better.

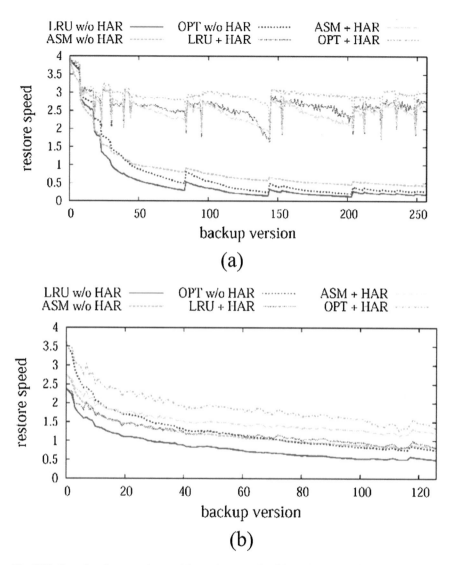

Fig. 8.10 Interplays between the rewriting and restore algorithms. (**a**) Kernel. (**b**) VMDK

8.3.7 Storage Cost

As discussed in index scheme, indexing 1 TB unique data of 4 KB chunks in DRAM, called *baseline*, costs $140, 57.14% of which is for DRAM. The cost is even higher if considering the high energy consumption of DRAM. The baseline storage costs are $0.23, $3.11, and $7.55 in Kernel, RDB, and VMDK respectively.

To reduce the storage cost, we either use HDD instead of DRAM for exact deduplication or index a part of fingerprints in DRAM for near-exact deduplication.

Table 8.4 The storage costs relative to the baseline which indexes all fingerprints in DRAM

Index/dataset	Fraction	Kernel	RDB	VMDK
EDPL/EDLL	DRAM	1.33%	1.40%	1.41%
	HDD	57.34%	55.15%	54.86%
	Total	58.67%	**56.55%**	56.27%
NDPL-64	DRAM	0.83%	0.83%	0.82%
	HDD	65.01%	61.25%	60.32%
	Total	65.84%	62.07%	61.14%
NDPL-128	DRAM	0.49%	0.48%	0.45%
	HDD	70.56%	66.08%	63.16%
	Total	71.04%	66.56%	63.61%
NDPL-256	DRAM	0.31%	0.31%	0.27%
	HDD	77.58%	73.58%	67.10%
	Total	77.89%	73.89%	67.36%
NDLL-64	DRAM	0.66%	0.70%	0.71%
	HDD	59.03%	55.27%	59.79%
	Total	59.69%	**55.97%**	60.49%
NDLL-128	DRAM	0.34%	0.35%	0.35%
	HDD	59.83%	55.34%	62.92%
	Total	60.17%	55.69%	63.27%
NDLL-256	DRAM	0.16%	0.17%	0.18%
	HDD	60.23%	55.65%	71.24%
	Total	60.39%	**55.82%**	71.42%

Table 8.4 shows the relative storage costs to the baseline in each dataset. NDPL-128 is NDPL of a 128:1 uniform sampling ratio. EDPL and EDLL have the identical storage cost, since they have the same deduplication ratio and key-value store. We assume that the key-value store in EDPL and EDLL is a database paired with a Bloom filter, hence 1 byte DRAM per stored chunk for a low false positive ratio. EDPL and EDLL reduce the storage cost by a factor of around 1.75. The fraction of the DRAM cost is 2.27-2.50%.

Near-exact deduplication of a high sampling ratio further reduces the DRAM cost, at a cost of decreasing deduplication ratio. Near-exact deduplication with a 128: 1 sampling ratio and 4 KB chunk size needs to achieve 97% of deduplication ratio of exact deduplication to avoid a cost increase. To evaluate this trade-off, we observe the storage costs of NDPL and NDLL under various sampling ratios. NDPL uses the uniform sampling, and NDLL is of the parameter (Top-4,4,1). As shown in Table 8.4, NDPL increases the storage cost in all datasets; NDLL increases the storage cost in most cases, except in RDB.

Finding (8): Although near-exact deduplication reduces the DRAM cost, it cannot reduce the total storage cost.

8.4 Design Recommendation

There is no one solution to achieve the best performance on these four performance indicators. Our goal is to find the design solution that can satisfy the following requirements: (1) lasting and stable high backup performance, (2) design trade-off on RAM usage, storage overheads, restore performance, and reliability.

According to the findings and suggestions, we summarize all four design solutions that meet the conditions for deduplication systems, including Recommended parameters and ad-vantages in Table 8.5. If the user needs the lowest storage cost, *EDLL* should be the best choice. Because *EDLL* has both a stable backup performance and the highest deduplication factor. It is recommended to use a random sampling method to reduce the write overheads of KV store. If the user needs a lower RAM usage, you can choose approximate deduplication, such as *NDPL* and *NDLL*. *NDPL* is easy to implement, and we recommend to use a uniform sampling method. *NDLL* has a more design parameter space and higher deduplication factor, and we recommend to use content-defined segmenting and similarity detection. If the user needs the stable restore performance, they can use rewriting algorithm to eliminate fragments. The fingerprint perfetching of EDPL has the least impact on the rewriting algorithm and it is recommended to use the *OPT* cache replacement algorithm. When the user needs a higher data reliability, it is recommended to use *DCT* replication to enhance the reliability of RAID5.

Table 8.5 Summary for all kinds of fingerprint index schemes with stable backup performance

Subspace	Recommended parameters	Advantages
	Key-value store on HDD	
EDLL	Content-defined segmenting	Low storage overheads
	Random sampling	
	RFAA restore algorithm	
	Key-Value store on RAM	Low RAM usage
NDPL	Uniform sampling	Simplest logical frame
	RFAA restore algorithm	
	Key-value store on RAM	
NDLL	Content-defined segmenting	
	Random or Min sampling method	Low RAM usage
	Similarity detection and prefetch	High deduplication ratio
	RFAA restore algorithm	
	Key-value store on HDD	High restore performance
EDPL	An efficient restore algorithm	Good interplay with rewriting
	OPT restore cache	

8.5 Concluding Remarks

This chapter describes the implementation of an open-source deduplication framework and discusses how to build a high-performance deduplication system with the optimized parameter configuration. Specifically, this chapter presents a general-purpose framework called DeFrame, which helps the readers to design different deduplication system solutions. This chapter performs lots of experiments based on different goals with N-dimensional parameter space for evaluation. Moreover, De-Frame is efficient to find reasonable trade-off among restore performance, memory overheads, and storage cost. Furthermore, this chapter presents some findings and gives some design recommendations for different goals to build an efficient deduplication system.

References

1. The Data Deluge. (2010). Retrieved from http://www.economist.com/node/15579717.
2. IDC. (2010). *The 2011 digital universe study*. Retrieved from http://www.emc.com/collateral/analyst-reports/idc-extracting-value-from-chaos-ar.pdf.
3. IDC. (2014). *The digital universe of opportunities: Rich data and the increasing value of the internet of things*. Retrieved from http://www.emc.com/leadership/digital-universe/2014 iview/executive-summary.htm.
4. Meyer, D. T., & Bolosky, W. J. (2011). A study of practical deduplication. In *Proceedings of the USENIX Conference on File and Storage Technologies (FAST)* (pp. 229–241).
5. El-Shimi, A., Kalach, R., Kumar, A., Ottean, A., Li, J., & Sengupta, S. (2012). Primary data deduplication—large scale study and system design. In *Proceedings of the USENIX Conference on Annual Technical Conference (ATC)* (pp. 285–296).
6. Wallace, G., Douglis, F., Qian, H., Shilane, P., Smaldone, S., Chamness, M., & Hsu, W. (2012). Characteristics of backup workloads in production systems. In *Proceedings of the USENIX Conference on File and Storage Technologies (FAST)* (Vol. 12, pp. 4-4).
7. Shilane, P., Huang, M., Wallace, G., & Hsu, W. (2012) WAN-optimized replication of backup datasets using stream-informed delta compression. In Proceedings of the ACM Transactions on Storage (ToS), 8(4), pp. 1-26
8. Amvrosiadis, G., & Bhadkamkar, M. (2015). Identifying trends in enterprise data protection systems. In *Proceedings of the USENIX Conference on Annual Technical Conference (ATC)* (pp. 151–164).
9. Li, K. (2004). *Merging technology: DD200 restorer*. Retrieved from http://storageconference.us/2004/Presentations/Panel/KaiLi.pdf.
10. Dubnicki, C., Gryz, L., Heldt, L., Kaczmarczyk, M., Kilian, W., Strzelczak, P., Szczepkowski, J., Ungureanu, C., & Welnicki, M. (2009). HYDRAstor: A scalable secondary storage. In *Proceedings of the USENIX Conference on File and Storage Technologies (FAST)* (Vol. 9, pp. 197–210).
11. HP (2014). *Eliminate the boundaries of traditional backup and archive*. Retrieved from http://www8.hp.com/us/en/products/data-storage/storage-backup-archive.html.
12. HYDRAstor-Scale-out Grid Storage Platform. (2014). Retrieved from http://www.necam.com/hydrastor/.
13. Commvault Simpana Software. (2014). Retrieved from https://www.commvault.com/simpana-software.
14. Bolosky, W. J., Douceur, J. R., Ely, D., & Theimer, M. (2000). Feasibility of a serverless distributed file system deployed on an existing set of desktop PCs. *Proceedings of the ACM SIGMETRICS Performance Evaluation Review, 28*(1), 34–43.

© Springer Nature Singapore Pte Ltd. 2022
D. Feng, *Data Deduplication for High Performance Storage System*,
https://doi.org/10.1007/978-981-19-0112-6

15. Policroniades, C., & Pratt, I. (2004). Alternatives for detecting redundancy in storage systems data. In *Proceedings of the USENIX Conference on Annual Technical Conference (ATC)* (pp. 73–86).
16. Zhu, B., Li, K., & Patterson, R. H. (2008). Avoiding the disk bottleneck in the data domain deduplication file system. In *Proceedings of the USENIX Conference on File and Storage Technologies (FAST)*, (Vol. 8, pp. 269–282).
17. Quinlan, S., & Dorward, S. (2002). Venti: A new approach to archival storage. In *Proceedings of the USENIX Conference on File and Storage Technologies (FAST)* (Vol. 2, pp. 89–101).
18. Muthitacharoen, A., Chen, B., & Mazieres, D. (2001). A low-bandwidth network file system. In *Proceedings of the Eighteenth ACM Symposium on Operating Systems Principles (SOSP)* (pp. 174–187).
19. Riggle, C. M., & McCarthy, S. G. (1998). Design of error correction systems for disk drives. *IEEE Transactions on Magnetics, 34*(4), 2362–2371.
20. Schroeder, B., & Gibson, G. A. (2007). Understanding disk failure rates: What does an MTTF of 1,000,000 hours mean to you? *Proceedings of the ACM Transactions on Storage (TOS), 3*(3), 8-es.
21. ZFS. Retrieved from http://en.wikipedia.org/wiki/ZFS.
22. Dropbox. Retrieved from http://www.dropbox.com/.
23. Guo, F., & Efstathopoulos, P. (2011). Building a high-performance deduplication system. In *Proceedings of the USENIX Conference on Annual Technical Conference (ATC)*.
24. Srinivasan, K., Bisson, T., Goodson, G. R., & Voruganti, K. (2012). iDedup: Latency-aware, inline data deduplication for primary storage. In *Proceedings of the USENIX Conference on File and Storage Technologies (FAST)* (Vol. 12, pp. 1–14).
25. Vrable, M., Savage, S., & Voelker, G. M. (2009). Cumulus: Filesystem backup to the cloud. *Proceedings of the ACM Transactions on Storage (ToS), 5*(4), 1–28.
26. Chen, F., Luo, T., & Zhang, X. (2011). CAFTL: A content-aware flash translation layer enhancing the lifespan of flash memory based solid state drives. In *Proceedings of the USENIX Conference on File and Storage Technologies (FAST)* (Vol. 11, pp. 77–90).
27. Du, Y., Zhang, Y., & Xiao, N. (2014). R-Dedup: Content aware redundancy management for SSD-based RAID systems. In *Proceedings of the IEEE Conference on International Conference on Parallel Processing (ICPP)* (pp. 111–120).
28. Spring, N. T., & Wetherall, D. (2000). A protocol-independent technique for eliminating redundant network traffic. In *Proceedings of the ACM SIGCOMM 2000 conference on Applications, Technologies, Architectures, and Protocols for Computer Communication* (pp. 87–95).
29. Agarwal, B., Akella, A., Anand, A., Balachandran, A., Chitnis, P., Muthukrishnan, C., Varghese, G. (2010). Endre: An end-system redundancy elimination service for enterprises. In *Proceedings of the USENIX Conference on Symposium on Networked System Design and Implementation (NSDI)* (pp. 419–432).
30. Waldspurger, C. A. (2002). Memory resource management in VMware ESX server. *Proceedings of the ACM SIGOPS Operating Systems Review, 36*(SI), 181–194.
31. Al-Kiswany, S., Subhraveti, D., Sarkar, P., & Ripeanu, M. (2011). VMFlock: Virtual machine co-migration for the cloud. In *Proceedings of the 20th International Symposium on High Performance Distributed Computing* (HPDC) (pp. 159–170).
32. Shilane, P., Wallace, G., Huang, M., & Hsu, W. (2012). Delta compressed and deduplicated storage using stream-informed locality. In *HotStorage*.
33. Xia, W., Jiang, H., Feng, D., & Tian, L. (2014). Combining deduplication and delta compression to achieve low-overhead data reduction on backup datasets. In *Proceedings of the IEEE Conference on Data Compression Conference (DCC)* (pp. 203–212).
34. Xia, W., Jiang, H., Feng, D., Tian, L., Fu, M., & Zhou, Y. (2014). Ddelta: A deduplication-inspired fast delta compression approach. *Performance Evaluation, 79*, 258–272.

35. Fu, M., Feng, D., Hua, Y., He, X., Chen, Z., Xia, W., Zhang, Y., & Tan, Y. (2015). Design tradeoffs for data deduplication performance in backup workloads. In *Proceedings of the USENIX Conference on File and Storage Technologies (FAST)* (pp. 331–344).

36. Mao, B., Jiang, H., Wu, S., & Tian, L. (2014). POD: POD: Performance oriented I/O deduplication for primary storage systems in the cloud. In *Proceedings of the 2014 IEEE 28th International Parallel and Distributed Processing Symposium (IPDPS)*, (pp. 767–776).

37. Koller, R., & Rangaswami, R. (2010). I/O deduplication: Utilizing content similarity to improve I/O performance. *Proceedings of the ACM Transactions on Storage (TOS), 6*(3), 1–26.

38. Opendedup. (2016). Retrieved from http://www.opendedup.org.

39. Drago, I., Mellia, M., Munafo, M. M., Sperotto, A., Sadre, R., & Pras, A. (2012). Inside dropbox: Understanding personal cloud storage services. In *Proceedings of the 2012 Internet Measurement Conference* (pp. 481–494).

40. Drago, I., Bocchi, E., Mellia, M., Slatman, H., & Pras, A. (2013). Benchmarking personal cloud storage. In *Proceedings of the ACM Conference on Internet Measurement Conference (IMC)*, (pp. 205–212).

41. Li, C., Shilane, P., Douglis, F., Shim, H., Smaldone, S., & Wallace, G. (2014). Nitro: A capacity-optimized SSD cache for primary storage. In *Proceedings of the USENIX Conference on Annual Technical Conference (ATC)* (pp. 501–512).

42. Im, S., & Shin, D. (2010). Flash-aware RAID techniques for dependable and high-performance flash memory SSD. *IEEE Transactions on Computers (TC), 60*(1), 80–92.

43. Chang, Y. B., & Chang, L. P. (2008). A self-balancing striping scheme for nand-flash storage systems. In *Proceedings of the 2008 ACM symposium on Applied Computing* (pp. 1715–1719).

44. Luo, X., & Shu, J. (2013). Load-balanced recovery schemes for single-disk failure in storage systems with any erasure code. In *Proceedings of the 2013 42nd International Conference on Parallel Processing (ICPP)* (pp. 552–561).

45. Anand, A., Muthukrishnan, C., Akella, A., & Ramjee, R. (2009). Redundancy in network traffic: findings and implications. In *Proceedings of the ACM SIGMETRICS*.

46. Pucha, H., Andersen, D. G., & Kaminsky, M. (2007). Exploiting similarity for multi-source downloads using file handprints. In *Proceedings of the USENIX Conference on Symposium on Networked System Design and Implementation (NSDI)*.

47. Sanadhya, S., Sivakumar, R., Kim, K. H., Congdon, P., Lakshmanan, S., & Singh, J. P. (2012). Asymmetric caching: Improved network deduplication for mobile devices. In *Proceedings of the 18th Annual International Conference on Mobile Computing and Networking (Mobi-Com)* (pp. 161–172)

48. Hua, Y., & Liu, X. (2012). Scheduling heterogeneous flows with delay-aware deduplication for avionics applications. *IEEE Transactions on Parallel and Distributed Systems, 23*(9), 1790–1802.

49. Gupta, D., Lee, S., Vrable, M., Savage, S., Snoeren, A. C., Varghese, G., Voelker, G. M., & Vahdat, A. (2008). Difference engine: Harnessing memory redundancy in virtual machines. In *Proceedings of the USENIX Conference on Operating Systems Design and Implementation (OSDI)*.

50. Miller, K., Franz, F., Rittinghaus, M., Hillenbrand, M., & Bellosa, F. (2013). XLH: More effective memory deduplication scanners through cross-layer hints. In *Proceedings of the USENIX Conference on Annual Technical Conference (ATC)* (pp. 279–290).

51. Jin, K., & Miller, E. L. (2009). The effectiveness of deduplication on virtual machine disk images. In *Proceedings of SYSTOR* (pp. 1–12).

52. Ren, J., & Yang, Q. (2010). A new buffer cache design exploiting both temporal and content localities. In *Proceedings of the 2010 IEEE 30th International Conference on Distributed Computing Systems (ICDCS)* (pp. 273–282).

53. Clements, A. T., Ahmad, I., Vilayannur, M., & Li, J. (2009). Decentralized deduplication in SAN cluster file systems. In *Proceedings of the USENIX Conference on Annual Technical Conference (ATC)* (Vol. 9, pp. 101–114).

54. Ng, C. H., Ma, M., Wong, T. Y., Lee, P. P., & Lui, J. C. (2011). Live deduplication storage of virtual machine images in an open-source cloud. In *Proceedings of the ACM/IFIP/USENIX International Conference on Distributed Systems Platforms and Open Distributed Processing* (pp. 81–100).

55. Zhang, X., Huo, Z., Ma, J., & Meng, D. (2010). Exploiting data deduplication to accelerate live virtual machine migration. In *Proceedings of the 2010 IEEE International Conference on Cluster Computing* (pp. 88–96).

56. Deshpande, U., Wang, X., & Gopalan, K. (2011). Live gang migration of virtual machines. In *Proceedings of the 20th International Symposium on High Performance Distributed Computing (HPDC)* (pp. 135–146).

57. Super Fast Hash. (2008). Retrieved from http://www.azillionmonkeys.com/qed/hash.html.

58. Nam, Y., Lu, G., Park, N., Xiao, W., & Du, D. H. (2011). Chunk fragmentation level: An effective indicator for read performance degradation in deduplication storage. In *Proceedings of the 2011 IEEE International Conference on High Performance Computing and Communications (HPCC)* (pp. 581–586).

59. Eshghi, K., & Tang, H. K. (2005). A framework for analyzing and improving content-based chunking algorithms. *Hewlett-Packard Labs Technical Report TR, 30.*

60. Schleimer, S., Wilkerson, D. S., & Aiken, A. (2003). Winnowing: local algorithms for document fingerprinting. In *Proceedings of the 2003 ACM SIGMOD International Conference on Management of Data* (pp. 76–85).

61. Rabin, M. O. (1981). *Fingerprinting by random polynomials.* Center for Research in Computing Techn., Aiken Computation Laboratory, Univ.

62. Xia, W., Jiang, H., Feng, D., Douglis, F., Shilane, P., Hua, Y., Fu, M., Zhang, Y., & Zhou, Y. (2016). A comprehensive study of the past, present, and future of data deduplication. *Proceedings of the IEEE, 104*(9), 1681–1710.

63. Bjørner, N., Blass, A., & Gurevich, Y. (2010). Content-dependent chunking for differential compression, the local maximum approach. *Journal of Computer and System Sciences, 76*(3–4), 154–203.

64. Zhang, Y., Jiang, H., Feng, D., Xia, W., Fu, M., Huang, F., & Zhou, Y. (2015). AE: An asymmetric extremum content defined chunking algorithm for fast and bandwidth-efficient data deduplication. In *Proceedings of the 2015 IEEE Conference on Computer Communications (INFOCOM)* (pp. 1337–1345). IEEE.

65. Zhang, Y., Feng, D., Jiang, H., Xia, W., Fu, M., Huang, F., & Zhou, Y. (2017). A fast asymmetric extremum content defined chunking algorithm for data deduplication in backup storage systems. *IEEE Transactions on Computers, 66*(2), 199–211.

66. Xia, W., Zhou, Y., Jiang, H., Feng, D., Hua, Y., Hu, Y., & Zhang, Y. (2016). Fastcdc: A fast and efficient content-defined chunking approach for data deduplication. In: *Proceedings of the USENIX Conference on Annual Technical Conference (ATC)* (pp. 101–114).

67. Kruus, E., Ungureanu, C., & Dubnicki, C. (2010). Bimodal content defined chunking for backup streams. In *Proceedings of the USENIX Conference on File and Storage Technologies (FAST)* (pp. 239–252).

68. Lillibridge, M., Eshghi, K., Bhagwat, D., Deolalikar, V., Trezis, G., & Camble, P. (2009). Sparse indexing: Large scale, inline deduplication using sampling and locality. In *Proceedings of the USENIX Conference on File and Storage Technologies (FAST)* (Vol. 9, pp. 111–123).

69. Min, J., Yoon, D., & Won, Y. (2010). Efficient deduplication techniques for modern backup operation. *IEEE Transactions on Computers, 60*(6), 824–840.

70. Romański, B., Heldt, Ł., Kilian, W., Lichota, K., & Dubnicki, C. (2011). Anchor-driven subchunk deduplication. In *Proceedings of the 4th Annual International Conference on Systems and Storage* (pp. 1–13).

71. Bhagwat, D., Eshghi, K., Long, D. D., & Lillibridge, M. (2009). Extreme binning: Scalables parallel deduplication for chunk-based file backup. In *Proceedings of the 2009 IEEE International Symposium on Modeling, Analysis & Simulation of Computer and Telecommunication Systems* (pp. 1–9).

72. Broder, A. Z. (1997). On the resemblance and containment of documents. In *Proceedings of the Compression and Complexity of SEQUENCES (Cat. No. 97TB100171)* (pp. 21–29).

73. Debnath, B. K., Sengupta, S., & Li, J. (2010). ChunkStash: Speeding up inline storage deduplication using flash memory. In *Proceedings of the USENIX Conference on Annual Technical Conference (ATC)* (pp. 1–16).

74. Agrawal, N., Bolosky, W. J., Douceur, J. R., & Lorch, J. R. (2007). A five-year study of file-system metadata. In *Proceedings of the USENIX Conference on File and Storage Technologies (FAST)*.

75. Bhagwat, D., Eshghi, K., & Mehra, P. (2007). Content-based document routing and index partitioning for scalable similarity-based searches in a large corpus. In *Proceedings of the 13th ACM SIGKDD International Conference on Knowledge Discovery and Data Mining* (pp. 105–112).

76. Tan, Y., Jiang, H., Feng, D., Tian, L., Yan, Z., & Zhou, G. (2010). SAM: A semantic-aware multi-tiered source de-duplication framework for cloud backup. In *Proceedings of the 2010 39th International Conference on Parallel Processing* (pp. 614–623).

77. Linux archive. *ftp://ftp.kernel.org*.

78. Xing, Y., Li, Z., & Dai, Y. (2010). Peerdedupe: Insights into the peer-assisted sampling deduplication. In *Proceedings of the 2010 IEEE Tenth International Conference on Peer-to-Peer Computing (P2P)* (pp. 1–10).

79. Wei, J., Jiang, H., Zhou, K., & Feng, D. (2010). MAD2: A scalable high-throughput exact deduplication approach for network backup services. In *Proceedings of the 2010 IEEE 26th Symposium on Mass Storage Systems and Technologies (MSST)* (pp. 1–14).

80. Pagh, R., & Rodler, F. F. (2004). Cuckoo hashing. *Journal of Algorithms, 51*(2), 122–144.

81. Nam, Y. J., Park, D., & Du, D. H. (2012). Assuring demanded read performance of data deduplication storage with backup datasets. In *Proceedings of the 2012 IEEE 20th International Symposium on Modeling, Analysis and Simulation of Computer and Telecommunication Systems (MASCOTS)* (pp. 201–208).

82. Kaczmarczyk, M., Barczynski, M., Kilian, W., & Dubnicki, C. (2012). Reducing impact of data fragmentation caused by in-line deduplication. In *Proceedings of the 5th Annual International Systems and Storage Conference* (pp. 1–12).

83. Lillibridge, M., Eshghi, K., & Bhagwat, D. (2013). Improving restore speed for backup systems that use inline chunk-based deduplication. In *Proceedings of the USENIX Conference on File and Storage Technologies (FAST)* (pp. 183–197).

84. Mao, B., Jiang, H., Wu, S., Fu, Y., & Tian, L. (2012). SAR: SSD assisted restore optimization for deduplication-based storage systems in the cloud. In *Proceedings of the 2012 IEEE Seventh International Conference on Networking, Architecture, and Storage (NAS)* (pp. 328–337).

85. SYMANTEC. (2010). *How to force a garbage collection of the deduplication folder*. Retrieved from http://www.symantec.com/business/support/index?page=content&id=TECH129151.

86. Belady, L. A. (1966). A study of replacement algorithms for a virtual-storage computer. *IBM Systems Journal, 5*(2), 78–101.

87. Linux Kernel. (2017). Retrieved from http://www.kernel.org/.

88. Tarasov, V., Mudrankit, A., Buik, W., Shilane, P., Kuenning, G., & Zadok, E. (2012). Generating realistic datasets for deduplication analysis. In *Proceedings of the USENIX Conference on Annual Technical Conference (ATC)* (pp. 261–272).

89. Gahm, J., & Mcknight, J. (2008). Medium-size business server & storage priorities. In *Enterprise Strategy Group*.

90. EMC Avamar. Retrieved from http://www.emc.com/avamar3.

91. Asigra Hybrid Cloud Backup and Recovery Software. Retrieved from http://www.asigra.com.

92. Commvault Simpana. Retrieved from http://www.commvault.com.
93. Rabin, M. O. (1981). Fingerprinting by random polynomials. In *Center for Research in Computing Techn*. Aiken Computation Laboratory, Univ.
94. NIST. (1993). Secure hash standard. In *FIPS PUB 180-1*.
95. Vrable, M., Savage, S., Voelker, G. M. (2009). Cumulus: Filesystem backup to the cloud. In *Proceedings of the USENIX Conference on File and Storage Technologies (FAST)*.
96. Bacula. Retrieved from http://www.bacula.org.
97. Amazon Simple Storage Service. Retrieved from http://aws.amazon.com/s3.
98. Harnik, D., Pinkas, B., & Shulman-Peleg, A. (2010). Side channels in cloud services: Deduplication in cloud storage. *IEEE Security & Privacy, 8*(6), 40–47.
99. Mulazzani, M., Schrittwieser, S., Leithner, M., Huber, M., & Weippl, E. (2011). Dark clouds on the horizon: Using cloud storage as attack vector and online slack space. In *Proceedings of the 20th USENIX Security Symposium* (pp. 1–11).
100. Dropbox API Utilities. (2011). Retrieved from https://github.com/driverdan/dropship.
101. Halevi, S., Harnik, D., Pinkas, B., & Shulman-Peleg, A. (2011). Proofs of ownership in remote storage systems. In *Proceedings of the 18th ACM conference on Computer and communications security (CCS)* (pp. 491–500).
102. Keelveedhi, S., Bellare, M., & Ristenpart, T. (2013). DupLESS: Server-aided encryption for deduplicated storage. In *Proceedings of the 22nd USENIX Security Symposium* (pp. 179–194).
103. Douceur, J. R., Adya, A., Bolosky, W. J., Simon, P., & Theimer, M. (2002). Reclaiming space from duplicate files in a serverless distributed file system. In *Proceedings of the 22nd International Conference on Distributed Computing Systems (ICDCS)* (pp. 617–624)
104. Bellare, M., Keelveedhi, S., & Ristenpart, T. (2013). Message-locked encryption and secure deduplication. In *Proceedings of Advances in Cryptology–EUROCRYPT* (pp. 296–312).
105. Bellare, M. & Keelveedhi, S. (2015). Interactive message-locked encryption and secure deduplication. In *Proceedings of IACR International Workshop on Public Key Cryptography (PKC)* (pp. 516–538).
106. Armknecht, F., Bohli, J. M., Karame, G. O., & Youssef, F. (2015). Transparent data deduplication in the cloud. In *Proceedings of the 22nd ACM SIGSAC Conference on Computer and Communications Security (CCS)* (pp. 886–900).
107. Naor, M., & Reingold, O. (2004). Number-theoretic constructions of efficient pseudo-random functions. *Journal of the ACM (JACM), 51*(2), 231–262.
108. Cox, L. P., Murray, C. D., & Noble, B. D. (2002). Pastiche: Making backup cheap and easy. *ACM SIGOPS Operating Systems Review, 36*(SI), 285–298.
109. Storer, M. W., Greenan, K., Long, D. D., & Miller, E. L. (2008). Secure data deduplication. In *Proceedings of the 4th ACM International Workshop on Storage Security and Survivability* (pp. 1–10).
110. Puzio, P., Molva, R., Önen, M., & Loureiro, S. (2013). ClouDedup: Secure deduplication with encrypted data for cloud storage. In *Proceedings of the 2013 IEEE 5th International Conference on Cloud Computing Technology and Science* (Vol. 1, pp. 363–370).
111. Li, J., Chen, X., Li, M., Li, J., Lee, P. P., & Lou, W. (2014). Secure deduplication with efficient and reliable convergent key management. *IEEE Transactions on Parallel and Distributed Systems, 25*(6), 1615–1625.
112. De Santis, A., & Masucci, B. (1999). Multiple ramp schemes. *IEEE Transactions on Information Theory, 45*(5), 1720–1728.
113. Shamir, A. (1979). How to share a secret. *Communications of the ACM, 22*(11), 612–613.
114. Duan, Y. (2014). Distributed key generation for encrypted deduplication: Achieving the strongest privacy. In *Proceedings of the 6th edition of the ACM Workshop on Cloud Computing Security (CCSW)* (pp. 57–68).
115. Liu, J., Asokan, N., & Pinkas, B. (2015). Secure deduplication of encrypted data without additional independent servers. In *Proceedings of the 22nd ACM SIGSAC Conference on Computer and Communications Security (CCS)* (pp. 874–885).

116. Xia, W., Jiang, H., Feng, D., & Hua, Y. (2011). SiLo: A similarity-locality based near-exact deduplication scheme with low RAM overhead and high throughput. In *Proceedings of the USENIX Conference on Annual Technical Conference (ATC)* (pp. 26–30).

117. FSL Traces and Snapshots Public Archive. (2014). Retrieved from http://tracer.filesystems. org/traces/fslhomes/2014/.

118. Broder, A. Z. (2000). Identifying and filtering near-duplicate documents. In *Annual symposium on combinatorial pattern matching* (pp. 1–10). Springer.

119. Jaccard, P. (1901). Etude de la distribution florale dans une portion des alpes et du jura. *Bulletin De La Societe Vaudoise Des Sciences Naturelles, 37*(142), 547–579.

120. MacDonald, J. (2000). *File system support for delta compression* (Masters Thesis). Department of Electrical Engineering and Computer Science, University of California at Berkeley.

121. Kulkarni, P., Douglis, F., LaVoie, J. D., & Tracey, J. M. (2004). Redundancy elimination within large collections of files. In *Proceedings of the USENIX Conference on Annual Technical Conference (ATC)* (pp. 59–72).

122. Douglis, F., & Iyengar, A. (2003). Application-specific Delta-encoding via resemblance detection. In *Proceedings of the USENIX Conference on Annual Technical Conference (ATC)* (pp. 113–126).

123. Jain, N., Dahlin, M., & Tewari, R. (2005). TAPER: Tiered approach for eliminating redundancy in replica synchronization. In *Proceedings of the USENIX Conference on File and Storage Technologies (FAST)* (Vol. 5, pp. 21-21).

124. Yang, Q., & Ren, J. (2011). I-CASH: Intelligently coupled array of SSD and HDD. In *Proceedings of the 2011 IEEE 17th International Symposium on High Performance Computer Architecture (HPCA)* (pp. 278–289).

125. Trendafilov, D., Memon, N., & Suel, T. (2002). *Zdelta: An efficient delta compression tool. Technical report*. Department of Computer and Information Science at Polytechnic University.

126. Broder, A. Z. (1993). Some applications of Rabin's fingerprinting method. In *Sequences II: Methods in communications, security, and computer science* (pp. 143–152). Springer.

127. Aronovich, L., Asher, R., Bachmat, E., Bitner, H., Hirsch, M., & Klein, S. T. (2009). The design of a similarity based deduplication system. In *Proceedings of SYSTOR 2009: The Israeli Experimental Systems Conference* (pp. 1–14).

128. Gailly, J., & Adler, M. (1991). *The gzip compressor*. Retrieved from http://www.gzip.org/.

129. GNU Software Archive. Retrieved from http://ftp.gnu.org/gnu/.

130. VMs Archives. Retrieved from http://www.thoughtpolice.co.uk/vmware//.

131. Redis. (2017). Retrieved from http://redis.io/.

132. Scilab Archives. Retrieved from http://www.scilab.org/.

133. Tarasov, V., Jain, D., Kuenning, G., Mandal, S., Palanisami, K., Shilane, P., Trehan, S., & Zadok, E. (2014). Dmdedup: Device mapper target for data deduplication. In *Proceedings of the 2014 Ottawa Linux Symposium (OSL)* (pp. 1–13).

134. Bhatotia, P., Rodrigues, R., & Verma, A. (2012) Shredder: GPU-accelerated incremental storage and computation. In *Proceedings of the 10th USENIX conference on File and Storage Technologies (FAST)* (Vol. 14, pp. 14).

135. Intel Solid-State Drive DC s3700 Series. (2014). Retrieved from http://www.intel.com/content/www/us/en/solid-state-drives/solid-state-drives-dc-s3700-series.html.

136. Huang, P., Wu, G., He, X., Xiao, W. (2014). An aggressive worn-out flash block management scheme to alleviate SSD performance degradation. In *Proceedings of the Ninth European Conference on Computer Systems (EuroSys)* (pp. 1–14).

137. Meister, D., Brinkmann, A., & Süß, T. (2013). File recipe compression in data deduplication systems. In *Proceedings of the 11th USENIX Conference on File and Storage Technologies (FAST)* (pp. 175–182).

138. MySQL. (2017). Retrieved from http://www.mysql.com/.

139. Fu, M., Feng, D., Hua, Y., He, X., Chen, Z., Xia, W., Huang, F., & Liu, Q. (2014). Accelerating restore and garbage collection in deduplication-based backup systems via

exploiting historical information. In *Proceedings of the 2014 USENIX Annual Technical Conference (ATC)* (pp. 181–192).

140. Strzelczak, P., Adamczyk, E., Herman-Izycka, U., Sakowicz, J., Slusarczyk, L., Wrona, J., & Dubnicki, C. (2013). Concurrent deletion in a distributed content-addressable storage system with global deduplication. In *Proceedings of the 11th USENIX Conference on File and Storage Technologies (FAST)* (pp. 161–174).

141. Anand, A., Muthukrishnan, C., Kappes, S., Akella, A., & Nath, S. (2010). Cheap and large CAMs for high performance data-intensive networked systems. In *Proceedings of the USENIX Conference on Symposium on Networked Systems Design And Implementation (NSDI)* (Vol. 10, pp. 29-29).

142. Lim, H., Fan, B., Andersen, D. G., & Kaminsky, M. (2011). SILT: A memory-efficient, high-performance key-value store. In *Proceedings of the Twenty-Third ACM Symposium on Operating Systems Principles (SOSP)* (pp. 1–13).

143. Debnath, B., Sengupta, S., & Li, J. (2011). SkimpyStash: RAM space skimpy key-value store on flash-based storage. In *Proceedings of the 2011 ACM SIGMOD International Conference on Management of Data* (pp. 25–36).

144. Zhou, Y.,Feng, D.,Xia W., et al. (2015). SecDep: A user-aware efficient fine-grained secure deduplication scheme with multi-level key management. In *Proceedings of 2015 31st Symposium on Mass Storage Systems and Technologies (MSST)* (pp. 1–14).

145. Tan Y., Jiang H., Feng D., et al. (2011). CABdedupe: A causality-based deduplication performance booster for cloud backup services. In *Proceedings of IEEE International Parallel Distributed Processing Symposium (IPDPS'11)* (pp. 1266–1277).

146. Zhang, Y., Xia, W., Feng, D., Jiang, H., Hua, Y., & Wang, Q. (2019). Finesse: Fine-grained feature locality based fast resemblance detection for post-deduplication delta compression. In *Proceedings of the 17th USENIX Conference on File and Storage Technologies (FAST)* (pp. 121–128).

147. Hu, Z., Zou, X., Xia, W., Jin, S., Tao, D., Liu, Y., Zhang, W., & Zhang, Z. (2020). Delta-DNN: Efficiently compressing deep neural networks via exploiting floats similarity. In *Proceedings of the 49th International Conference on Parallel Processing (ICPP)* (pp. 1–12).

Printed in the United States
by Baker & Taylor Publisher Services